Designing Sustainable Prosperity

Designing Sustainable Prosperity

Natural Resource Management for Resilient Regions

Edited by

Doris Hiam-Galvez
Hatch Ltd
Designing Sustainable Prosperity
Vancouver, British Columbia
Canada

Published by John Wiley & Sons, Inc., Hoboken, New Jersey.
Published simultaneously in Canada.

For general information on our other products and services or for technical support, please contact our Customer Care Department within the United States at (800) 762-2974, outside the United States at (317) 572-3993 or fax (317) 572-4002.

Wiley also publishes its books in a variety of electronic formats. Some content that appears in print may not be available in electronic formats. For more information about Wiley products, visit our web site at www.wiley.com.

Library of Congress Cataloging-in-Publication Data Applied for
Paperback ISBN: 9781394253296

Cover Design: Wiley
Cover Images: © szefei/Shutterstock, © Phil Seu Photography/Getty Images

Set in 9.5/12.5pt STIXTwoText by Straive, Pondicherry, India

SKY10078834_070324

To my husband John and my son Kamir, who have inspired me to think in terms of ecosystems and supported me every step of the way as I developed this new approach to sustainable development.

To Hatch, an environment of "entrepreneurs with a technical soul", where the impossible becomes possible, which continuously pushed us towards creating positive change.

To John Bianchini and Joe Lombard, thank you for your trust. Your belief in the importance of creating a better world for generations to come has been both a guiding light and a source of motivation.

Special thanks to the hummingbird, whose visits brought me joy, energy, and resilience to bring this concept to life.

To all contributors: Allan Moss, whose deep knowledge of the industry has been indispensable; Richard Blewett for his geoscience expertise that has greatly enriched our work; and Lily So, whose passion for sustainability and unwavering commitment carried this book to fruition. To the younger contributors, whose enthusiasm inspired us all.

To my clients and colleagues, who are friends in this vast family of the industry. Your collective spirit and support have been invaluable.

I am immensely proud and deeply grateful to be part of this community, striving together towards a brighter world.

Contents

List of Contributors

Richard Blewett
GeoSystems Consulting, Geoscience
Royalla, New South Wales
Australia

John Hiam
TechMat Consulting, Metallurgical
Mechanical Engineering, Climate
Vancouver, British Columbia
Canada

Doris Hiam-Galvez
Hatch Ltd, Designing Sustainable
Prosperity
Vancouver, British Columbia
Canada

Anthea Kong
Hatch Ltd, Industrial Clean Technologies
Mississauga, Ontario
Canada

Brittany MacKinnon
Hatch Ltd, Pyrometallurgy
Brisbane, Queensland
Australia

Allan Moss
Sonal Mining Technology Inc.
Vancouver, British Columbia
Canada

Lily Lai Chi So
Hatch Ltd, Innovation and Investments
Mississauga, Ontario
Canada

Matthew Tutty
Hatch Ltd, Climate Change
Mississauga, Ontario
Canada

Jake Wyman
Hatch Ltd, Renewable Power and Electrical
Distribution
Mississauga, Ontario
Canada

Preface: My Journey into Sustainability

In a world where the future can often appear daunting and uncertain, I offer a message of hope and a clear path forward. We stand at a critical juncture during an energy transition that demands a surge in metal production. My belief is that we can meet this rising demand while upholding principles of sustainability.

With three decades of experience, I have had the privilege of reshaping corporate organisations, driving innovation and expanding businesses across four continents. My journey has been driven by a commitment to reshape our relationship with the planet.

Today, I present a vision: the transformation of the metals industry into a driving force for a global positive change. Consider the extractive industry not merely as a source of materials but as a potent catalyst with the potential to nurture resilient, thriving regions that benefit our environment and communities alike.

Now, more than ever, this industry has the potential to lead a transformative shift towards sustainability. It extends beyond resource extraction: it is about redefining our role as stewards of the Earth, becoming an integral and harmonious part of the interconnected ecosystem.

Within the pages of this book, I introduce an alternative approach, a path that requires a fresh perspective on how we perceive our connection to the world. It demands a shift towards systems thinking, and within these chapters, I offer a structured multidisciplinary approach that dares to question existing assumptions. This new approach sets the stage to create the conditions for an emergency of sustainability. It is time to liberate ourselves from established and outdated paradigms and embrace innovation. This innovation seeks to balance economic expansion with environmental stewardship.

As we embark on this transformative journey, let us tap into our collective creativity, envisage a brighter future and redefine our legacy. This book serves as a call to action for the industry to become a cornerstone in the global sustainability movement. Let us not merely dream of a better tomorrow; let us take concrete steps to make it a reality. Together, we have the power to shape a future where our industry not only supplies resources but also shines as a beacon of sustainable practices, environmental stewardship and healthy and prosperous communities. Together, we can turn challenges into opportunities, forging a prosperous, sustainable world for generations to come.

Endorsements

As a perennial advocate and speaker for the Minerals Industry and its role in society, I have long spoken on how our industry makes life possible on our planet. In that same context, I have had the great privilege to view *Designing Sustainable Prosperity* as working draft. And what a treat that opportunity has turned out to be. Doris and her collaborators have brought together a broad and deep menagerie of facts, models and case studies demonstrating the role and future potential the minerals industry has in terms of making a world a better place. And at the same time, they have helped describe how all the pieces fit together in a simple and practical way. Not only is her work a great collection of ideas and references, but also a great story in terms of how it brings the pieces together and guides how we can each make a difference.

As the leader at Anglo American, we brought together 100,000 people in a conversation to define our Purpose, to 'Re-imagine mining to improve people's lives'. In *Designing Sustainable Prosperity,* Doris and her team are imagining what real change looks like and how it can be delivered.

Mark Cutifani, Chairman Vale Base Metals, Former CEO at Anglo American plc.

The world's population of 8 billion inhabitants requires the resource industry, not only to supply its material needs, but also to be part of the disciplined solution in improving the environmental condition of the planet. This requires a shift in how resource extraction is undertaken, to ensure there are sympathetic sustainable local outcomes. The mining industry is a guest in the areas where it works, and the local social, regulatory and political frameworks need to support and encourage timely and efficient mineral extraction, if we as a society are to meet legislated aspirations for 2050 and beyond.

With her extensive industry perspective, Doris Hiam-Galvez has set out a roadmap to achieve resource demand through *Designing Sustainable Prosperity*. Her book is evidenced-based and recommended for Corporations, Politicians and Regulators to understand the sequencing necessary to access the energy transition metals and realise 2050 aspirations in a sustainable manner.

Robert Quartermain, DSc, Canadian Mining Hall of Fame Inductee 2022, Co-Chair Dakota Gold Corp, Former Executive Chairman Pretium Resources Inc.

In her timely new book, Doris Hiam-Galvez challenges communities that are dependent on traditional mining and other extractive industries to adopt a 'Designing Sustainable Prosperity' planning approach that can produce vibrant and diversified local economies. Hiam-Galvez makes a richly illustrated pitch for communities to imagine and embrace a future in which their historical, single-dimensional industries leverage tech and other advances to set them up for durable, long-term success – while simultaneously attracting new, sustainable businesses into their region.

Communities around the world increasingly know that climate change and related water, biodiversity and land use impacts require new ways of thinking about – and prospering in – a more durable and sustainable future. This book offers a must-read road map for these communities. It presents an optimistic, 'bottom up'" collaboration recipe that leavens outside expertise with community-based history, capabilities and ambition to move in new directions and create reimagined local economies that are strong, sustainable and well prepared for the long haul.

David J. Hayes, Professor at Stanford University, formerly a senior White House climate advisor for President Biden and the Deputy Secretary and Chief Operating Officer of the U.S. Department of the Interior for Presidents Obama and Clinton.

A dichotomy exists between the growing need for minerals to sustain modern society and facilitate global decarbonisation on the one hand, and the negative perception of the mining industry on the other. DSP applies a systems approach to unlocking the long-term potential of resource-rich regions by integrating mining operations with regional economic development, environmental sustainability and community well-being. It acknowledges the finite life of mining projects and proposes a practical methodology to seek holistic solutions to ensure the ultimate diversification and growth of the regional economy well beyond the closure of the mine.

This book describes a thought-provoking pathway by which mining can become a catalyst for the collective sustainable good through the integration of the social, economic, technical and environmental dimensions that connect a mine to its host region.

John MacKenzie, CEO and Founder of Capstone Copper.

Communities often underappreciate the potential for resource projects to be a force for good. More often, new project proposals are met with the language of fear and opposition. A different mindset using our collective creativity would instead see such opportunities as a potential catalyst for achieving a larger and more sustainable future. By reshaping our relationships, the language could be one of hope and vision.

Designing Sustainable Prosperity is a must-read for those visionary community and regional leaders that want to leave a positive legacy for their future generations.

David Harquail, Chair of the Board of Directors at Franco-Nevada, Board Director at the Bank of Montreal.

The linked climate and biodiversity crises are challenging natural resource communities. We need to protect and restore biodiversity, support local development, extract minerals for the energy transition, and deploy wind, solar, pipelines and transmission infrastructure. Without new strategies and tools, communities and our environment will suffer. *Designing Sustainable Prosperity* offers an innovation at the right time. Hiam-Galvez starts with the right question – what does prosperity look like for this region and this community. This new way of thinking is underpinned by an engineer's attention to planning and detail. It is a must-read for those that care about development of our natural resources, from town halls and Tribal councils to mining company boardrooms.

Stephen D'Esposito, President and CEO at Regeneration and Resolve.

Doris has written a most thoughtful book that is most valuable to the well-being of communities in mitigating the impacts of development on their lives. This is a particularly invaluable resource for the Indigenous communities of the world who are often the nearest neighbours to major natural resource development projects. All Indigenous leaders committed to sustainable development should read this book. It details a practical method for balancing development with environmental protection and enhancing quality of life.

Calvin Helin, Bestselling Author, Entrepreneur, Speaker, Indigenous Leader.

Today, the world is being challenged as never before to find a way forward that is positive for people and ecosystems, over the short term and long, and using processes that fairly distribute the benefits, costs, risks, responsibilities and accountabilities. It is only by doing so that today's generations will not undermine the ability of future generations to meet their needs. In a nutshell, this is the essence of practical implementation of sustainability concepts. *Designing Sustainable Prosperity* takes up the challenge of practical application of sustainability concepts in a constructive and thought-provoking way. This book is essential reading for those who wish to address this tough challenge head-on.

Designing Sustainable Prosperity addresses the sustainability challenge in a constructive and thought-provoking way. It is essential reading for those wishing to address this tough challenge.

R. Anthony Hodge, Ph.D., P.Eng., Adjunct Professor, Queen's University (Canada) and University of Queensland (Australia), President, International Council on Mining and Metals, 2008-2015.

This book is essential for our time. We have been living OFF the land for far too long. Now, we need to live WITH the land. The primary change we need to make is perceptual. We need to understand more deeply that we are not outside observers of Nature, but literally extensions of Nature. How Nature goes, so do we. Understanding this relationship is essential for transformation. Also essential is to translate this understanding into action. And this is where Doris succeeds with a clear outline of how communities can collaborate with business to create a sustainable future for both!

Beau Lotto, Professor of Neuroscience, Author, Founder and CEO of several start-ups.

As the energy transition is accelerating, the world is moving from fossil fuels to metals as one of its primary resources. A new approach is needed however to ensure that the extractive mindset is accompanied by stewardship of the Earth and our own environment. This book provides a unique and comprehensive framework to balance mankind's resource needs with environmental stewardship, by design. Doris Hiam-Galvez proposes to deliberately and systematically *Design Sustainable Prosperity* (DSP) based on understanding regional, societal, economic, environmental and core values. The book encourages systems thinking to break the current piece-meal approaches by leveraging a society's creativity to create sustainable regions. A step-wise process calls for key parties to sit around the table and follow such deliberate design in a staged process, laid out eloquently in 10 chapters backed up by real-world examples. The writer comes with 30 years of experience in the extractive industry, and her passion for doing it different clearly shines through the pages of a book that is a must-read for countries' leadership, or anyone with a drive to make the twenty-first century a better one.

Jef Caers, Founder Mineral-X, Stanford University.

Doris Hiam-Galvez and her co-authors have tackled a problem that has been the curse of mining for decades, if not centuries, the ability to generate lasting value in a region that hosts a mine. *Designing Sustainable Prosperity* (DSP) is a novel, organised and systems-based approach that provides a template for seeking sustainable economic activity from mining that continues after the mine has closed. The basis for this approach starts with our understanding of the earth, the three-dimensional geoscientific data that explains where concentrations of metals occur and how the metals can be recovered responsibly with minimum impact on the environment. This fundamental connection to the earth is rarely evaluated comprehensively when considering long-term sustainability. As is well recognised, resource development requires extensive consultation with communities, experts and numerous stakeholders and this book explains how this can be done at various stages throughout the process. The book provides a new approach that deserves serious consideration, and at the very least, it should increase efforts to determine how mining can contribute more completely and effectively to regional prosperity.

John Thompson, CIO at Regeneration, Board Director at KoBold and MineSense, Honorary Professor at the University of Bristol, Adjunct Professor at the University of British Columbia and Cornell University.

Redesigning societies, leveraging and valuing all resources, is the challenge of the day: more than a hopeful vision, DSP charts a viable, evidence-based path to realise this opportunity. A must-read for mining and minerals leaders, policy-shapers and stakeholders who want to play a role in reshaping linkages between a complex industry and its environment and invite others to join them on the journey of collaborative design. From rethinking education as the spark igniting the fire of change, to demonstrating the value unlocked by DSP for people and their natural environment through several high-level case studies, Doris Hiam-Galvez and her co-authors offer insights and inspiration, connect the dots across disciplines and expectations, challenge the status quo and energise us to reach new heights.

Ludivine Wouters, Managing Partner, Latitude Five.

1

Executive Summary

Doris Hiam-Galvez

Hatch Ltd, Designing Sustainable Prosperity, Vancouver, British Columbia, Canada

Sustainability is the responsible use of resources, including preserving the environment, to meet current needs without compromising the needs of future generations. Designing Sustainable Prosperity (DSP) introduces a systems thinking approach to unlock resource rich regions while placing sustainability and community integration at its core. It seeks to balance economic development, environmental sustainability and community wellbeing, thereby fostering the emergence of resilient and thriving regions. It harnesses the extractive industry as a catalyst for lasting prosperity.

The risks facing resource rich regions include climate change, reliance on a commodity economy and a shortage of skilled labor, as illustrated in Figure 1.1. DSP tackles these challenges with a holistic approach, unlocking untapped potential for system wide solutions that address climate change. It pioneers innovative solutions to diversify and scale the economy, positioning the region as leaders in the knowledge economy. Once the focus is identified, the education system is adapted to provide skills needed for sustainable prosperity, as illustrated in Figure 1.2.

Figure 1.1 Regional Risks.

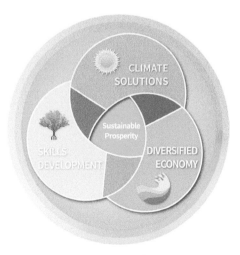

Figure 1.2 DSP Mitigates those Risks.

DSP fosters long-term regional prosperity by promoting sustainable economic diversification and treating regions as interconnected ecosystems. As shown in Figure 1.3, regional economic activity is plotted on the vertical axis and time on the horizontal axis. The red curve represents the economic vulnerability of these resource rich regions. It shows an increase in economic activity during operations followed by a decline as resources are depleted. The yellow curve illustrates how support industries, follow a similar trend. The preferred outcome is shown by the green curve, to sustain a stable liveable level of economic activity that extends beyond the life of extractive operations. This represents a big shift from short-term to long-term thinking.

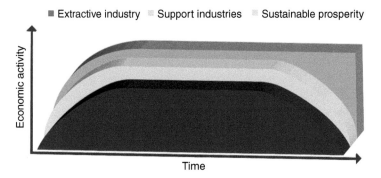

Figure 1.3 Designing Sustainable Prosperity (DSP). *Source:* Adapted from Hiam-Galvez, Prescott, & Hiam (2020).

The DSP process comprises three key planning phases:

1) Building the Design Inputs: Unlocking the hidden potential of natural resources and human talent.
2) Designing System Solutions: Key parties collaborate in a seven-step process (see Figure 1.4).
3) Preparing for Implementation: Creating an environment for effective and purposeful execution.

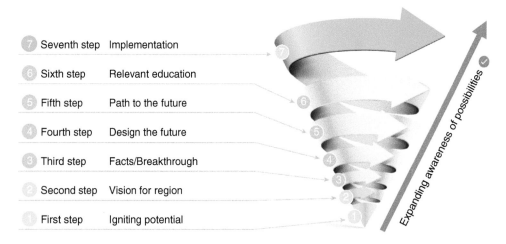

Figure 1.4 The seven steps.

The outcomes are investment packages, business cases for key enterprises and strategic roadmaps.

DSP is a facilitated process designed to unlock the full potential of both human and natural resources to co-create solutions for the region. Figure 1.4 depicts the seven steps of DSP, emphasizing the gradual expansion of possible opportunities and actionable steps towards sustainable prosperity. The first two steps focus on preparing and empowering the team so that they are open to collaborate and ready to embark on this journey. When the team is prepared, they move into the planning phase (steps 3 to 6). The final step focuses on implementation readiness and establishing the environment for sustained action and progress.

United Nations Sustainable Development Goals (UN SDGs) indicators are used to establish a baseline assessment of the region's economic, environmental, and social conditions. During the design phase, business cases are strategically prioritized based on these indicators, evaluating their potential regional impact and alignment with global development goals. After implementation, these indicators continue to serve as a critical tool, measuring success and monitoring progress towards achieving a sustainable and equitable future for the region.

High Level Case Studies

Southern Peru

In *Southern Peru*, DSP harnesses the region's abundant solar potential, showcasing how the region's expertise in complex processing for copper production can be adapted to future sustainable industries. With current capabilities, the region can seize the opportunity to lead in sustainable energy solutions, not just by acquiring technologies, but by showcasing its ability to adapt and evolve complex technologies. Their opportunity lies in advancing direct solar seawater desalination technology by concentrated solar power (CSP): a complex thermal process. This approach addresses water scarcity, promotes precision agriculture and encourages the development of value added food products.

Ring of Fire

In Canada's Northern Ontario *Ring of Fire*, a crucial ecosystem thrives above mineral deposits. This development opportunity centres on the preservation and well-being of the wetland ecosystem. The region holds the potential to establish itself as a leading hub for peatland regeneration and carbon capture expertise, a transformation catalysed by responsible resource extraction and guided by Indigenous knowledge. Collaborative partnerships with research centres further enable the delicate balance between ecological preservation and sustainable socio-economic development, firmly rooted in peatland regeneration.

Quebec

Quebec, Canada, is a hub for hydropower energy, and faces energy scarcity as it works towards its decarbonisation goals by 2050. DSP focuses on establishing a stable grid supply, integrating various green energy sources including hydropower, wind, solar and hydrogen.

The expertise in managing energy demand and the supply of a complex green energy ecosystem through the seasons via innovative digital solutions provides a basis for a knowledge industry. Decarbonisation goals can be met while diversifying the economy.

Northwest British Columbia

In *Northwest British Columbia (NWBC)*, Canada, DSP proposed a plan to accelerate mining development while protecting the environment and respecting Indigenous cultures in the region. The process involves leveraging NWBC's abundant natural resources supported by good infrastructure. The initiative focuses on a collaborative mining economy emphasising shared resources and remote operation of mines. This region, with its robust hydropower supply, is uniquely positioned to evolve into a low carbon solutions hub. By integrating its strength in hydropower with innovative approaches in remote sensing, the region could set a benchmark in environmentally sustainable practices.

Reference

Hiam-Galvez, D., Prescott, F., & Hiam, J. (2020). Designing Sustainable Prosperity "DSP": A collaborative effort to build resilience in mining producing regions. *CIM Journal*, 11(1):69–79.

2

Introduction

Doris Hiam-Galvez

Hatch Ltd, Designing Sustainable Prosperity, Vancouver, British Columbia, Canada

In the face of global challenges such as climate change, biodiversity loss, food insecurity, poverty and scarcity of essential skills, the role of metals becomes increasingly important. These challenges necessitate a responsible and rapid increase in metal supply, especially in the context of the energy transition. Our mission extends beyond meeting energy and infrastructure needs; it encompasses the protection of our environment and the enhancement of the global quality of life (World Science Forum, 2022).

We find ourselves at a pivotal moment in history, poised for an extraordinary opportunity to lead radical change that benefits future generations. Key focus areas include:

1) Redefining exploration and extraction for minimal environmental harm and embracing the circular economy with an unwavering focus on value creation.
2) Adapting the education system and reskilling the workforce for the future low carbon economy.
3) Forging partnerships with multidisciplinary experts, investors and communities to co-create nature positive solutions.
4) Designing systems that minimise waste and ideally contribute to environmental enrichment.
5) Rethinking the role of geoscience. DSP places a strong emphasis on unlocking regional subsurface potential, drawing from the knowledge of the earth's formation to explore new possibilities. This approach advocates for the widespread integration of geoscience in shaping our society's decisions. Stewart and like-minded individuals also champion geoscience's evolving role in addressing society challenges (Gloaguen, et al., 2022; Stewart, 2023; Wadsworth, Llewellin, Brown, & Aplin, 2020).

Envisage a future where the extractive industry, driven by visionary and committed leadership, embraces transformative change. In this future, the extractive industry transforms into a catalyst for sustainable regional prosperity. It cultivates an environment conducive to innovation and empowers communities, effectively eliminating the dependency where the extractive industry serves as the sole provider of income and essential services. This approach paves the way for diverse, resilient economies.

DSP facilitates a shift from an extractive-centred business focus to diversified economies, ensuring sustainable employment post-extraction. It anticipates the skills needed for the future, promoting education and workforce development.

While numerous publications have explored economic diversification in resource rich regions, this book complements existing literature because it details how to implement those strategies. These specifics, often overlooked in previous works on sustainable development or resource-based economic diversification, are vital for achieving lasting regional resilience and sustainability.

Existing publications on diversifying economies in resource rich regions primarily offer policy recommendations and insights into the challenges and opportunities faced by developing countries in leveraging their extractive industries for inclusive sustainable development.

There is a growing commitment to maximise the benefits that extractive industries bring to their host countries. T. Addison and A. Roe in their book 'Extractive Industries: The Management of Resources as a Driver for Sustainable Development' published by Oxford Press in 2018 and featuring contributions from international experts across various disciplines provides a broader array of ideas and recommendations in critical policy areas. It underscores the importance of emphasising the development role of extractive industries and explores how actions on climate change will shape the sector's future (Addison & Roe, 2018).

Since the publication of that book in 2018, there has been significant progress with respect to the extractive industry's efforts to benefit their host countries, albeit with varying results. Many of these diversified economies built in resource rich regions remain heavily reliant on extractive investment and may not survive when these industries reach the end of their life cycle. The focus of the extractive industry to date has primarily been on the short term benefits.

The DSP method goes beyond today's thinking and enables the diversification away from the extractive industry in a way that is sustainable after the metal resources have been depleted. DSP is a complement to economic planning, sustainable development and diversification in resource rich regions. The DSP method is a transformative approach on how to do it in a practical and constructive manner integrating multiple disciplines and leveraging a system solution. This book provides comprehensive guidance on how to initiate and sustain a series of activities resulting in a resilient and sustainable region.

Furthermore, this book serves as a valuable complement to Mark Carney's 'Value(s): Building a Better World for All' (Carney, 2021). Carney advocates for long term thinking and argues that by re-evaluating and reshaping our values, we can build a world that prioritises the well-being of all and the planet.

Kate Raworth's 'Doughnut Economics' has gained recognition as a holistic framework for sustainable development (Raworth, 2017). It considers the interconnectedness of economic, social and environmental issues to create a thriving society within the ecological boundaries of the planet.

Karl-Henrik Robert's (Robert, 1997) 'The Natural Step' offers a sustainability framework that provides a roadmap for transitioning towards a sustainable future. It serves as a framework for navigating the complexities of sustainability and working towards a future where human well-being is supported within ecological limits of the planet.

DSP supports and complements the ideas put forward by these authors by adding the crucial dimension of 'how to implement'.

DSP is also aligned with the United Nations (UN) Sustainable Development Goals (SDGs) (United Nations Department of Economic and Social Affairs, 2015), strategically prioritising business cases that promise the highest impact. This alignment serves as the backbone for fostering sustainable regional economies and securing long term employment prospects. The DSP process delivers investment packages, crafting business cases for leading enterprises and charting roadmaps for scaling impactful solutions. Furthermore, it adapts the education system to cultivate the necessary skills and foster an environment that nurtures creativity and innovation.

At its core, DSP centres on bolstering the local community's ability to create value whilst addressing future unmet market needs. The strategy revolves around unleashing both human and natural resource potential, fostering skill development and facilitating economic diversification that is sustainable. It starts by empowering individuals, enabling them to broaden their horizons and explore new possibilities. The use of earth and geoscience plays a crucial role in uncovering regional potential, with a special focus on tackling water challenges and securing green and abundant energy sources, the essential drivers of economic diversification.

This book goes beyond theory, featuring high level case studies that showcase the knowledge economy and identifies potential enterprises capable of driving sustainable progress. It provides an exploration of the method and its high level application across various regions, emphasising the transition towards innovative knowledge-based industries. The goal is to provide a catalyst for action.

Comprising 10 chapters, the book commences by highlighting the importance of metals, sustainability, prosperity beyond financial performance and the instrumental role of geoscience. These provide the basis for the DSP objectives. We then dive into the DSP method, highlighting specific case studies spanning regions in Australia, Peru and Canada, and discuss the critical adaptation of education systems to support future economies. There is also a chapter envisaging the future of society.

DSP is not merely an aspiration but a holistic approach, empowering individuals with a new way of thinking as they navigate the challenges of our time. It encompasses economic resilience, environmental sustainability, collaborative efforts and a commitment to enhancing the quality of life while considering long term consequences. Most notably, DSP primarily focuses on solutions that address climate change and promote nature positive initiatives as a system that leads to long term sustainability and prosperity for all. In its concluding section, this book serves as a rallying call to action, encouraging readers to become active contributors in shaping a sustainable and prosperous future. The book is a valuable resource for driving positive change and creating a prosperous, sustainable world, paving the way for transformative progress.

References

Addison, T., & Roe, A. (2018). *Extractive Industries: The Management of Resources as a Driver of Sustainable Development*. Oxford University Press.

Carney, M. (2021). *Value(s): Building a Better World for All*. McClelland & Steward.

Gloaguen, R., Ali, S. H., Herrington, R., Ajjabou, L., Downey, E., & Stewart, I. S. (2022). Mineral revolution for the wellbeing economy. *Global Sustainability*, 5: e15.

Raworth, K. (2017). *Doughnut Economics: Seven Ways to Think Like a 21st-Century Economist*. London: Random House.

Robert, K. -H. (1997). *Natural Step: A Framework*. Pegasus Communications, Inc.

Stewart, I. (2023). Geology for the wellbeing economy. *Nature Geoscience*, 16(2):106–107.

United Nations Department of Economic and Social Affairs. (2015). *17 Sustainable Development Goals (SDGs)*. Retrieved from Sustainable Development: https://sdgs.un.org/goals

Wadsworth, F., Llewellin, E., Brown, R., & Aplin, A. (2020, January 14). *Earth sciences face a crisis of sustainability*. Retrieved from Times Higher Education: https://www.timeshighereducation.com/opinion/earth-sciences-face-crisis-sustainability

World Science Forum. (2022, December 9). *Declaration of the World Science Forum 2022*. Retrieved from Declaration of the 10th World Science Forum on Science for Social Justice: https://worldscienceforum.org/contents/declaration-of-world-science-forum-2022-110144

3

Metals – A Key to the Future

John Hiam[1] and Doris Hiam-Galvez[2]

[1] TechMat Consulting, Metallurgical, Mechanical Engineering, Climate, Vancouver, British Columbia, Canada
[2] Hatch Ltd, Designing Sustainable Prosperity, Vancouver, British Columbia, Canada

Modern society cannot survive without metals, and building a sustainable future also depends on metals. The global energy transition hinges on an increased demand for metals. These essential materials are pivotal in driving the adoption of, for example, renewable energy (see Figure 3.1) and electric vehicles. But with this demand comes a crucial challenge: the long development timeline for metal mining projects, especially copper. It can take up to 30 years from exploration to production, half of which are often consumed by the permitting process.

Figure 3.1 Metals are key to sustainable energy production; for example, wind energy.
Source: JordiDelgado/iStockphoto.

Designing Sustainable Prosperity: Natural Resource Management for Resilient Regions,
First Edition. Edited by Doris Hiam-Galvez.

Most metals exist in the ground in the form of ores (some combination of elements as compounds). Even gold which is present as a pure metal exists primarily as an impurity as a very small percentage of the total bodies of minerals. So, in most cases, mining is followed by some degree of separation and then refining is required to obtain the desired pure metals.

This begins with mining which is defined by operations such as 'digging the stuff out of the ground' to obtain the ore which will typically be mixed with other material and thus require separating and putting into a form suitable for further processing to extract the metal. There are many opportunities to improve the efficiencies of the process steps and to thus reduce energy consumption, reduce waste and find uses for that waste.

For future protection of the environment and to conserve our precious resources, huge efforts are needed to modify the systems just described. Major efforts are underway to improve mining efficiencies, reduce energy consumption, reduce waste, develop uses for waste materials and make similar changes to metal production and applications. Replacement of rare, expensive or 'difficult to produce' metals is also being investigated.

Reimagining Mining

The process of extraction of metals includes mining and processing to produce a concentrate that is transported from the mine site to industrial complexes such as smelters, refineries and steel mills to obtain semi-finished products such as copper wire, steel or aluminium strip or bars, etc. While mines are located where the ores are found in nature, industrial plants tend to be located close to major population centres for logistical efficiency.

Looking at the future, the mining industry must adopt a more holistic view of ore bodies. This means moving beyond the traditional single commodity focus to recognise and use the full array of resources an ore body offers. Such a comprehensive approach promises to revolutionise mining operations, maximising resource utilisation and drastically reducing environmental impacts.

There is a need to reimagine mining as an integrated system that transcends the traditional silos of geology, mining and processing. This perspective requires a thorough understanding of ore bodies, focusing on value creation. The mining industry, by embracing innovative and sustainable practices, will not only accelerate the transformation but also align with the global sustainability goals. The industry can position itself as a key player in the journey towards a sustainable future.

In Situ Recovery – A Possible Alternative Mining Method

The future of mining beckons a transformative approach, one where we extract only what is valuable, mirroring the efficiency and harmony of nature. In situ recovery (ISR), while heralding a new era in mining, comes with its own set of challenges. With technological advances, this technology is poised to redefine mining, focusing on extracting value while leaving the host rock in place.

ISR is an alternative method to extract valuable minerals without physically mining the rock, so no waste rock and no tailings. This avoids the capital- and energy-intensive recovery methods of conventional mining and reduces surface disturbances associated with waste rock piles and tailings facilities.

Mining is traditionally done by digging a big hole in the ground and separating the valuable minerals from the waste. With ISR, there is no need to dig a big hole but instead, dissolve the valuable minerals underground and pump the solution up to the surface.

The technology has been applied successfully over the last several decades for commercial mining of potash because it is soluble in water. Also acid-soluble minerals such as uranium ores are hosted in porous rocks. It has also been applied for a few shallow highly fractured copper oxides in the 1980s and 1990s with mixed results. There are ores in similar physical conditions that can be amenable to ISR. The challenge is with unfractured rocks hosted in hard rock minerals deep underground such as copper sulphides. The challenge is that the rock permeability and porosity must be enhanced to expose the minerals to the leaching solution (Hiam-Galvez, Gerber, Pekrul, & Earley, 2020).

The commercial success of ISR has been supported by advances in fragmentation methods, mapping of the ore body, hydrological modelling, containment of fugitive lixiviant and monitoring groundwater quality (Hiam-Galvez, Harding, Krawchuk, Slabbert, & Brown, 2018).

ISR is a promising technology but with challenges. Reducing the costs while achieving fragmentation and effective containment to yield good recovery is key. Chances of success depend on the characteristics of each ore body. So, significant development work is needed to expand its application to the more complex ore bodies.

Integrating DSP into Mining Projects

Before embarking on any regional development initiatives, a DSP evaluation should take place as a precursor to project commencement, allowing for a comprehensive examination of the entire ecosystem, an appreciation of the possibilities and all the components of the system.

By engaging with key parties, this process aims to unlock the full spectrum of possibilities. Armed with an understanding of what is achievable and how it fits into the broader system, subsequent projects can then commence.

This strategic approach ensures that the focus remains on the region's long term prosperity, facilitating informed decision-making throughout project phases from inception to completion.

Moreover, this approach can be implemented throughout the project phases, from exploration to completion, as described below and illustrated in Figure 3.2:

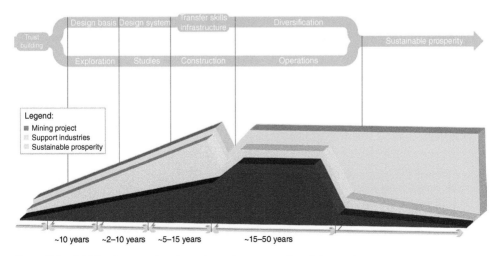

Figure 3.2 DSP implementation throughout the stages of a mining project.

1) **Exploration Stage: Uncovering Subsurface Potential**
   - Assess the Subsurface Profile: Beyond identifying potential ore bodies, exploration could delve into the complete geoscience profile of the region. This includes other mineral deposits, sources of groundwater, geothermal energy and storage opportunities.
   - Understanding the environmental, social and economic implications of mining activities and how they fit into the larger ecosystem of the region. Explore opportunities for conservation and diversification.

2) **Conceptual Stage: Envisaging**
   - Further understanding of the subsurface potential in terms of new sources of energy, water and potential storage capabilities.
   - Align with regional development goals, ensuring it contributes to a broader economic and environmental vision. Focus on understanding the broader ecosystem and identifying how mining can complement and enhance regional prosperity through circular economy practices.
   - Develop an integrated vision where mining operations are harmonised with regional sustainable development goals. This includes integrating circular economy concepts like waste reduction and resource sharing, ensuring mining activities contribute positively to the local environment and community.

3) **Feasibility Stage: DSP Planning**
   - At this stage, a comprehensive DSP study should be completed to identify sustainable expansion avenues beyond mining, preparing for a diversified economy that attracts new investments and sustains economic activity after construction and mining. The mining investor becomes a 'catalyst' and is instrumental in creating an environment to attract new investors to co-create a resilient and sustainable region and plan to maintain some level of economic activity post-construction.

4) **Construction Stage: Catalyst for the Implementation of DSP Short term Actions**
   - Use construction activities to set the foundation for a circular, resilient regional economy. This could involve using sustainable materials, supporting local circular enterprises and establishing infrastructure that can be repurposed post-mining. Training programs should focus on skills relevant to both mining and the circular economy and how to expand those skills to regional resilience. Effectively train and absorb some of the construction workforce and economic activity of construction into other emerging enterprises that would have been attracted to the region because of the DSP work.

5) **Operational Stage: Actively Catalyse Partnerships**
   - Operate the mine within a sustainable framework, ensuring minimal environmental impact. Simultaneously, advance DSP initiatives to contribute as a catalyst to inspire the emergency of new enterprises, preparing for eventual reduction in mining-related activities and workforce. Via DSP, actively catalyse partnerships and initiatives that reinforce the region's economic diversification, embedding mining operations within a larger, resilient and sustainable ecosystem. Continually explore innovations to enrich the larger ecosystem.

6) **Post-mining Transition Stage: Gradual Transition to a Diversified Economy**
   - Strategically plan for a gradual transition from mining to a resilient diversified economy. Capitalise on the DSP progress made during early stages to ensure new

enterprises are operational, effectively absorbing the workforce and economic activity from mining. The aim is to avoid boom-and-bust cycles post-construction and post-mining.

As illustrated in Figure 3.2, the creation of a sustainable prosperity for the region serves as a key driver from the very onset of the project, even preceding exploration. The key lies in embedding key parties from the project inception. Even before exploration, efforts towards DSP can be initiated by collaborating with local communities in diversification activities to start building trust. During the exploration phase, comprehensive geological and ecological information is gathered to assess not only the commercial potential of mineral deposits but also the broader potential of the region for diversification.

Throughout these stages, mining is seen as an integral part of a larger system contributing to regional prosperity. By executing a well-planned DSP strategy, mining can transform from a standalone operation to a key player in a sustainable, resilient and diversified economic ecosystem that benefits all.

Metal Production and Applications

The discussion will be split into the following topics:

1) Steel production
2) Manufacturing of goods such as automobiles, buildings, batteries, phones, computers, etc.
3) Designing for recyclability
4) Selection decisions based on issues such as recyclability, the lowest possible waste products, pollution and other environmental impacts, resource consumption, life cycle costs, impact, etc.

Steel Production

Steelmaking is a carbon-intensive activity, and breakthrough technologies are being developed to address the steel industry's aspirations for carbon neutrality by 2050. As the industry pursues the goal of sustainable steelmaking, more efficient operation of existing infrastructure will be critical to achieving the emissions reductions targets set by national and international policy initiatives in the near term.

The traditional steel production route involves charging the prepared iron ore to a blast furnace together with coke (purified coal) and fluxes such as limestone to reduce the iron ore (containing iron oxides) to iron. This results in liquid iron saturated with carbon which is then removed in the next operation by blowing oxygen through the liquid steel to oxidise the carbon which is emitted as carbon dioxide. All these process steps and subsequent ones are very expensive and result in huge quantities of greenhouse gases (GHGs). New ways of improving these existing operations are being developed; however, a transformation of the steelmaking production route is required for decarbonisation as described in the climate change section. Similar approaches are being considered for many other metals, such as aluminium, which requires tremendous quantities of energy to produce the metal.

Manufacturing of Goods and Design for Recyclability

Figure 3.3 shows one specific application: the automobile. Many materials are used in addition to metals and so many processes are involved in making the materials and then processing them for assembly into the vehicle. For example, the engine uses aluminium alloys, steel alloys and lubricants. Batteries to produce power to start the vehicles, for example, and, in the case of electric vehicles, to power electric motors to propel the cars, use lead, manganese, nickel, cobalt, lithium and other chemicals. The metals in the electric motors for electric vehicles (EVs) include rare earth metals. The whole process of producing cars involves the energy consumed to manufacture them, transport materials, components and finished vehicles and pollution from processes such as painting, treating and handling of waste such as scrap and effluents. Then there is the energy consumed by use of the vehicles, emissions such as CO_2 and further consumption of materials to keep the vehicles maintained. At the end of the life of the car, it is scrapped and some of the materials are recovered for recycling while the remaining materials are scrapped.

Figure 3.3 Metals play a big role in our transition to sustainable electric vehicles. *Source:* iStockphoto/Getty Images.

In addition to the direct consumption of materials and energy, there is the associated infrastructure needed to accommodate the movement, parking, transportation of the supplies for production and use and scrapping of the vehicles. The same general issues apply to other products such as phones, buildings, appliances such as refrigerators, furniture and computers.

Work is being carried out to further develop recycling of metals and materials, thus reducing the need to extract them from the earth. Examples of these efforts include separating the individual metals employed in mobile phones and reusing EV batteries as storage units after they have been removed from the EVs at the end of their useful lives in the cars. A goal for the automotive industry is for the whole car to be recyclable. BMW is one manufacturer often cited as being close to this target. As previously mentioned, EV batteries and electric motors require metals not previously used in motor vehicles (nickel, cobalt, lithium and rare earth metals). They present significant environmental, availability and political problems associated with their procurement and production. For example, lithium, if obtained from South American salt flats, requires large amounts of water to extract in a region with severe water shortages. Many of the other metals also have production and environmental issues with their production. Most are present as minor constituents in the ore bodies, are therefore expensive to separate and large amounts of waste material are generated in their production. For these reasons, research is being carried out to find replacements.

Because of all the issues with procurement, production and environmental concerns, recycling is being pursued wherever possible. Therefore, to facilitate recycling, there are now many efforts to include recyclability in the design process for all products.

During the DSP process, many of the activities described above could be considered for potential business opportunities.

References

Hiam-Galvez, D., Gerber, E., Pekrul, J., & Earley, D. (2020). In Situ Recovery (ISR) – The Permitting Challenge. *ALTA 2020 In-Situ Recovery Conference* (pp. 64–73). Melbourne: ALTA Metallurgical Services Publications.

Hiam-Galvez, D., Harding, D., Krawchuk, P., Slabbert, W., & Brown, R. (2018). An Alternative Mining Concept. *Canadian Institution of Mining, Metallurgy and Petroleum*. Vancouver.

4

Sustainability – Climate Change and Environment
John Hiam[1] and Doris Hiam-Galvez[2]

[1] *TechMat Consulting, Metallurgical, Mechanical Engineering, Climate, Vancouver, British Columbia, Canada*
[2] *Hatch Ltd, Designing Sustainable Prosperity, Vancouver, British Columbia, Canada*

Climate Change

James Lovelock, a British scientist, developed a concept he called the Gaia hypothesis. It describes the earth (see Figure 4.1) as being like a living organism that lives in equilibrium. Any changes to the conditions will be countered by other changes to maintain the equilibrium. This is a good description of the current state of the Earth, which has had a stable climate for a very long time.

Figure 4.1 The Blue Marble, an image of Earth taken from space by the Apollo 17 crew in 1972. *Source:* NASA (2007) / Public domain / https://visibleearth.nasa.gov/images/55418/the-blue-marble-from-apollo-17/55420l.

However, since the beginning of the industrial revolution, the temperature of the Earth's atmosphere has begun to rise, the rate of rise increasing especially over the past two decades (Mann, 2021). The primary cause of this increase has been shown to be the consequence of human activities resulting in greenhouse gas (GHG) emissions such as carbon dioxide (CO_2) and methane (CH_4) into the atmosphere (United States Environmental Protection Agency, 2022).

Since CO_2 has long been known to increase the greenhouse effect (heating of the atmosphere from heat radiating from the ground), the rising levels of CO_2 and other GHG emissions from industrial, agricultural, domestic heating and other activities are now known to be contributing to global warming. The impact of this phenomenon is increased frequency of extreme weather events such as storms and hurricanes, flooding of coastal and low-lying land, flora and fauna species loss because of their inability to adapt to the unaccustomed rapid changes, sea level rise from ice melting and water warming and the associated coastal flooding, disruption to food production, increases in tropical and other diseases and so on. There is also an increase in the acidity of the oceans resulting from the increase in CO_2 absorbed from the atmosphere. This is causing loss of coral reefs and many of the small creatures such as plankton, krill, etc. There also may be a tipping point when the temperature rise triggers the release of more GHGs such as methane from places such as the arctic permafrost and initiates a positive feedback loop. Six of the nine planetary boundaries put forward by the Stockholm Resilience Centre have already been crossed (see Figure 4.2).

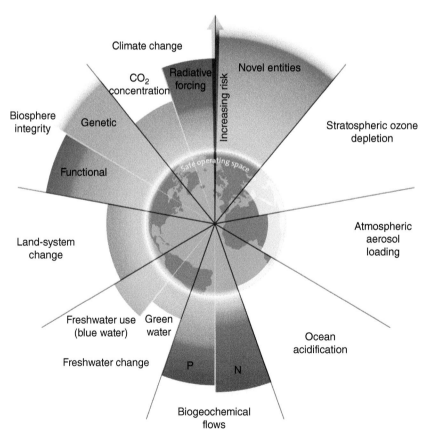

Figure 4.2 The planetary boundaries concept, which presents nine processes that regulate the stability and resilience of the Earth system. *Source:* Stockholm Resilience Centre (2023) / with permission of Stockholm Resilience Centre.

The combination of all these factors is leading to increasing efforts to halt the climate changes and avoid escalating to a catastrophe in the future. Three basic approaches are being taken, namely reduce (ideally eliminate) GHGs from activities, if not possible, capture the emissions, or develop methods of removing them from the atmosphere.

To eliminate emissions, so-called 'clean energy' generation is being developed with significant progress already having been made, especially in efficiency and cost. Examples are solar, wind, hydroelectric power generation, geothermal, tidal and nuclear. None of them are totally clean because of the equipment manufacturing required for all of them, environmental disruption and waste produced, especially waste disposal and storage for nuclear power. All need metals.

To reduce emissions, transportation developments include hybrid and battery electric cars and trucks, fuel cell powered cars and buses, improvements to internal combustion engine efficiencies and increasing mass public transportation to replace individual car use. Industrial companies are looking at ways to improve efficiency which reduces emissions and replaces processes with more environmentally friendly alternatives. Steel manufacturing companies, for example, may be able to eliminate the use of coke and the GHG generator known as a blast furnace and the oxygen steel making process that removes carbon from blast furnace iron. Many newer steel companies use scrap metal as their raw material which they melt in electric arc furnaces (EAFs) and avoid the coke oven, blast furnace and oxygen steel making steps entirely (Fan & Friedmann, 2021). Emerging processing routes such as direct reduction of iron ore to metallised iron using hydrogen are also being developed. New ways of making concrete (a major source of CO_2) are being explored. Insulation of commercial buildings and houses is also an option for reducing emissions by lowering the need for energy to heat or cool. Companies and research organisations are also exploring ways to capture CO_2 being emitted by operations. Chemically combining CO_2 with other compounds to produce solid or liquid products or pumping the CO_2 underground are examples.

Lastly, there are many efforts to develop new technology to remove CO_2 from the atmosphere. Trees are very efficient at removing CO_2 from the air. So, planting vast numbers of them is being contemplated. Artificial 'trees' are being developed to mimic the photosynthesis that they use (Lackner, 2014).

In addition to adoption of new technologies to reduce emissions, carbon taxes are being introduced in many countries. Activities that produce CO_2 are taxed based on the quantity of emissions which raises costs to the purchasers of the products or services and thus creates disincentives. Although sometimes controversial, the concept has been found to be a successful way of reducing the generation of CO_2.

There are many opportunities for new companies to emerge to develop, improve and commercialise the technologies and products that address the need to tackle climate change. These opportunities should be considered when setting up sustainable economies.

Environment

Natural resources are everywhere, many in remote locations where local communities depend on the land. We need to learn how to extract the required resources with net positive or zero or minimum impact on the environment. And to repair the environment where we have adversely changed it. Change should be benevolent.

For most of its more than four billion years' lifetime, the earth has evolved without the influence of humans and ultimately reached a period of stability. All life on earth is interconnected from the simplest single cell organisms to the most complex creatures such as humans. Damaging or eliminating specific species can result in chain reactions to many other life forms including disruption to the food chain, even right up to humanity. Lovelock's living organism is no longer in equilibrium but rather in a state of flux.

When humans first evolved, they lived in harmony with nature, sustaining themselves by being 'hunters and gatherers'. Then, humans discovered the benefits of agriculture and raising livestock and thus began to manipulate the environment, probably initially in the area known as the fertile crescent between the Tigris and Euphrates rivers of Mesopotamia. One of the best examples of the effects of this new lifestyle can be seen in Western Europe and the British Isles where over the centuries, the countryside has been completely transformed from its original mainly forested state to the present manicured fields and occasional small remaining forests. Most of the original wildlife and forested areas have disappeared.

In the nineteenth century, the beginning of the Industrial Revolution accelerated the transformation of the landscape and the environment affecting the flora and fauna and the atmosphere. The rate of change (almost all of it detrimental to the environment) has been increasing with population growth, the rising intensity of industrialisation and development of advanced agricultural technology. Examples of the effects of all this activity and what is being done, including future actions that are needed to remedy the situation, are briefly described in the following discussion. Modern agriculture often entails growing the same crops year after year on the same land and application of very aggressive pesticides resulting in soil deterioration and possible catastrophic events such as the early twentieth century dustbowl drought in the United States of America. The transformation of natural land for growing crops also resulted in destruction of the local wildlife and the associated natural habitat needed for both the resident animals and those that migrate through. Genetic engineering of crops to resist insects resulted in elimination of those 'pests' which wiped out vital food sources for migrating and resident birds and, thus, led to their subsequent demise. Large herds of grazing animals such as sheep and cattle have also drastically changed the landscape. Water consumption has been a key need for all agricultural activities frequently resulting in depriving the natural flora and fauna and causing subsequent environmental damage and species loss.

As the human population continues to grow, urbanisation of the land has also increased, resulting in 'concreting over' of that land and thus still more environmental destruction, habitat loss and species extinction.

Industrial activities including mining, manufacturing of cars and many other goods and building highways, bridges and cities and all the associated infrastructure have led to extensive damage to the environment. Mining, both surface and underground, produces large volumes of waste as the desired constituents in the product are separated out, thus producing so-called tailings which then must be stored or processed in some way, potentially polluting the surrounding countryside. There are also effluents that may be discharged into the environment and, in the past, were often untreated resulting in still more environmental pollution. Water consumption by mining usually results in extensive rerouting and potential contamination unless extensive post-use repurification is undertaken. This is also true of many other industrial activities and the ongoing use by cities.

Wherever humans live or work, the by-products of their activities include vast quantities of waste (garbage), much of which is discarded by shipping it to nearby dump sites or to other locations further away, sometimes even to other countries. One type of waste that has become a serious environmental problem is plastic (Organization for Economic Co-operation and Development, 2022). Plastic waste in the ocean is particularly damaging to the life of the inhabiting creatures, eventually contaminating the seafood that we consume. Whatever form of waste we consider, if it is dumped, then it is contaminating and destroying the environment. Modern civilisation has been called the 'throwaway society' because of our propensity to discard our purchases when we grow tired of them even if they are still functional. The clothing and automotive industries are examples of this behaviour. Such behaviour exacerbates the waste problem.

Powering all the activities of humans requires vast amounts of energy which needs to be generated and has resulted in many different approaches which have various levels of impact on the environment. Coal was the primary source of energy for the industrial revolution, but it has resulted in severe environmental damage ranging from mining and the associated disturbing of the immediate locale to copious amounts of dirty air pollution, GHG emissions and severe health problems for the workers producing the coal and for the general population. Other forms of fossil fuel such as oil and natural gas used in electricity generation and as the energy source for various forms of transportation also contribute to environmental damage from the massive oil sands destruction to smog from combustion of oil to GHG emissions. Hydroelectric power generation, despite being thought of as 'green' can result in loss of land through flooding behind the dams constructed to hold back water. Downstream areas are also affected by altered or reduced flows. Nuclear power requires the disposal of the radioactive waste products of the fission process. They can remain radioactive for thousands of years. So safe storage is a challenge that has to be overcome.

Since we cannot continue to treat the environment as we have been and since the resources we are consuming are not limitless, we must change. The following summarises some of the approaches that are already being undertaken along with others that are being considered, most of them offering opportunities for future business ventures. The general philosophy underlying the approaches includes improving efficiency and reducing or eliminating waste, recycling materials, capturing polluting by-products and setting worldwide standards to ensure remedies are adopted by everyone without loss of competitiveness resulting from rogue companies and countries ignoring the new standards. This will also involve developing new technologies and behaviours that are environmentally friendly. The challenge is to do all of this while raising the standard of living and quality of life for all people on the planet.

The agricultural industry is developing ways of using water more efficiently, adopting practices that do not degrade soil quality, developing more environmentally friendly treatments to protect crops from pests and finding ways to grow organic crops more efficiently. The general public must be educated on the benefits of decreasing consumption of 'animal-based' food and moving to more 'plant-based' diets which will significantly reduce energy consumption and GHG emissions. A huge proportion of food generated is never consumed. So, ways of ensuring this no longer happens must be found. This overall situation has created opportunities for new and existing businesses to develop more appealing and nutritious foods to encourage the diet changes needed and for development of more sustainable agricultural technology, which also addresses the need for adequate quantities of food for everyone.

Land management must be significantly increased involving activities such as protection of existing and frequently sensitive areas, planting of trees, creating larger areas of natural terrain to stop the rampant species extinction currently occurring and controlling the increasing urban sprawl that is destroying the environment. Business and therefore new employment opportunities will arise as such efforts are increased.

The mining industry has recognised the need for reducing its environmental impact and many initiatives are already underway. Improving efficiency by automation and adopting/developing new process technology is part of this effort. Improving water management including treatment to eliminate pollution and waste management including finding uses for the waste instead of just storing it as well as capturing fugitive emissions are also necessary. Ways of restoring the land to its former state or some other environmentally friendly condition should be part of any mining project. In some cases, perhaps underground instead of open pit mining can be utilised.

A major source of environmental deterioration is the production and use of energy from fossil fuels and many activities are already underway to address the problem. Alternative generation technologies such as solar and wind power are becoming more viable through technological advances and cost reductions so that they will soon be cost-competitive with traditional sources. Coal has been considered the lowest cost fuel but the high societal costs of the damage to people's health and the environment are not normally accounted for. The fossil fuel industry still benefits from large governmental subsidies which mask the true cost of using those resources. This situation needs to be addressed. The new green energy sources will continue to need subsidising in the form of initial support for technological research and development; however, they must eventually reach commercially viable status. Fossil fuels will still be needed in significant quantities for many more years while the new processes increase their market share.

There are still many opportunities to reduce the pollution they cause and to achieve efficiencies in their use. For example, 'clean coal' development is being explored. It must be remembered that oil is also used for applications other than fuel, for example, plastics and other materials. So, the demand for oil will continue. As with the other topics, there are many opportunities for the development and growth of existing and new industries focussed on technology, efficiency improvements and pollution elimination.

Cities and the transportation systems within them and between them are major sources of pollution, waste and environmental degradation. Energy consumption by transportation maintains the demand for fossil fuels, the detrimental effects being noted in the previous paragraphs. Transportation is also a significant contributor to GHGs. Hence, the development of various forms of propulsion technologies such as battery electric, fuel cell powered and hybrid electric vehicles. New concepts in city design to reduce the need for personal vehicles, development of comprehensive public transportation, new approaches such as vehicle sharing, retrofitting of buildings and new construction techniques are all aimed at energy efficiency improvements. Cities also generate large amounts of waste that, as previously mentioned, must be disposed of. Recycling is already being attempted but much more needs to be done to reduce the volume of unusable material including designing products for ease of recycling. There will be major new employment opportunities in many of these fields in existing and new companies.

Over the last few decades there has been a dramatic increase in the use of existing and new plastics. Unfortunately, many of the products or goods incorporating plastics are discarded and dumped into the environment after use. This has resulted in massive amounts of pollution in the oceans and on land. Most of the plastics remain in the environment for a very long time but frequently degrade into microplastic particles that are then consumed by creatures and consequently enter the food chain. The long term effect of this is not yet known but it is certainly not thought to be beneficial. The material that does not degrade or has not had time to do so can also arrive in the oceans or in the land environment and be mistakenly eaten by animals and trapping fish or other larger animals leading to their death. To avoid these problems, plastics that quickly biodegrade after use need to be developed and recyclable plastics be developed and used. In addition, methods of ensuring plastics do not enter the environment after use and methods of removing those that are already contaminating the environment need to be developed.

In summary, significant progress in reducing the negative environmental impact of human activity has been achieved but much more needs to be done. We already have a lot of the required technology but not all of it has been applied and there is a need for much more development. Governments worldwide should be involved in providing business environment that stimulates sustainable activities and discourages undesirable ones. The DSP process considers all the issues relevant to the environment including those discussed here and identifies the relevant and appropriate business opportunities.

References

Fan, Z., & Friedmann, S. J. (2021). Low-carbon production of iron and steel: Technology options, economic assessment, and policy. *Joule*, 5(4): 829–862.

Lackner, K. S. (2014). The Use of Artificial Trees. In R. E. Hester, & R. M. Harrison, *Issues in Environmental Science and Technology No. 38: Geoengineering of the Climate System* (pp. 80–104). Cambridge: The Royal Society of Chemistry.

Mann, M. E. (2021). Beyond the hockey stick: Climate lessons from the common era. *Proceedings of the National Academy of Sciences of the United States of America*, 18(39):e2112797118.

NASA. (2007, November 30). *Blue Marble – Image of the Earth from Apollo 17*. Retrieved from https://www.nasa.gov/content/blue-marble-image-of-the-earth-from-apollo-17/

Organization for Economic Co-operation and Development. (2022, February 22). *Plastic pollution is growing relentlessly as waste management and recycling fall short, says OECD.* Retrieved from https://www.oecd.org/environment/plastic-pollution-is-growing-relentlessly-as-waste-management-and-recycling-fall-short.htm

Stockholm Resilience Centre. (2023). *Stockholm Resilience Centre*. Retrieved from Planetary boundaries: https://www.stockholmresilience.org/research/planetary-boundaries.html

United States Environmental Protection Agency. (2022, May 5). *Understanding Global Warming Potentials*. Retrieved from https://www.epa.gov/ghgemissions/understanding-global-warming-potentials

5

Prosperity Beyond Financial Performance

Allan Moss[1], Brittany MacKinnon[2], and Doris Hiam-Galvez[3]

[1] Sonal Mining Technology Inc., Vancouver, British Columbia, Canada
[2] Hatch Ltd, Pyrometallurgy, Brisbane, Queensland, Australia
[3] Hatch Ltd, Designing Sustainable Prosperity, Vancouver, British Columbia, Canada

What is Prosperity?

Prosperity is frequently equated with financial success, a perspective challenged by notable figures such as Mark Carney. In his book 'Value(s): Building a Better World for All', Carney advocates for a paradigm shift from society's focus on financial metrics to broader values. World-renowned thought leaders in strategic management, Michael Porter and Mark Kramer, have also urged for a reinvention of prosperity, away from short term financial performance, and shifting towards true value creation.

While businesses often prioritize financial metrics, assessing the social and environmental impacts in resource rich areas remains a complex challenge. DSP addresses this gap by striving for the long term wellbeing of regions, integrating water, energy, biodiversity, food, health, education and wealth beyond the boundaries of the extractive industry. This holistic view underscores the need to consider sustainable prosperity and community wellbeing from the outset of the investment rather than as an afterthought.

What Happens After the Extractive Industry?

Typically, the extractive industry operates under a system that measures outcomes primarily by financial returns to stakeholders, an approach intensified by market demands for short term gains. Communities share some of the short term benefits of the industry in expectations of immediate rewards such as employment and the provision of services such as utilities and health care. The question arises as to the legacy of these operations and how the region will benefit in the longer term. The emerging issues that determine future developments are community integration, sustainability and regional prosperity beyond the extractive industry investments.

Within the metals industry, more attention has been focused on sustainability (Carvalho, 2017). Various researchers have created specific sustainability indicators for the mining and metals industry (Azapagic, 2004). For example, the United Nations Sustainable Development Goals' (UN SDGs') indicators are being used to propose impact and environmental assessment for mining activities (Dehghani, Bascompta, Khajevandi, & Farnia, 2023; Han, Cao, & Yan, 2021).

Why UN SDGs for Sustainable Development?

The alignment with sustainable development has emerged as a key focus, resulting in the development of various frameworks (van Tulder, 2018). The UN SDGs have become the primary framework shaping the global development agenda until 2030.

The move towards the SDGs represents a significant paradigm shift in how we understand sustainable development and the roles of entities like corporations. The new sustainable development paradigm encapsulated by the SDGs addresses the need for active engagement by companies and other societal actors in building a resilient world.

The SDGs present:

- a shift from the linear economy, characterised by wasteful production systems to the circular economy to potentially reach a waste free economy.
- transitioning from exclusive production models to an inclusive economy, catering to all segments of society.
- embracing a sharing economy, promoting collaboration, shared stewardship and sustainable market practices.
- an evolution towards prosperity beyond financial performance.

To this end, DSP aligns itself with the SDGs. These 17 interlinked goals, adopted by the United Nations in 2015 as a part of the 2030 Agenda for Sustainable Development, provide a comprehensive blueprint. They aim to create a better and more sustainable future for all, tackling global challenges such as poverty, inequality, climate change, environmental degradation, peace and justice (United Nations Department of Economic and Social Affairs, 2015).

DSP embraces and applies the SDGs fostering positive change in sustainable regional development that enhances the well-being of both current and future generations.

Why is DSP Aligned with the UN SDGs?

The SDGs serve as a universally accepted blueprint for sustainable development. DSP aligns with the SDGs to empower collective efforts and resources for maximum positive impact on regions. Figure 5.1 illustrates DSP's alignment with the SDGs emphasising key values: water, energy, biodiversity, food, health, education and wealth.

Furthermore, the SDGs offer a comprehensive and holistic approach to development, recognising the interconnectedness of social, economic and environmental dimensions. This integrated perspective emphasises the importance of multifaceted solutions. For instance,

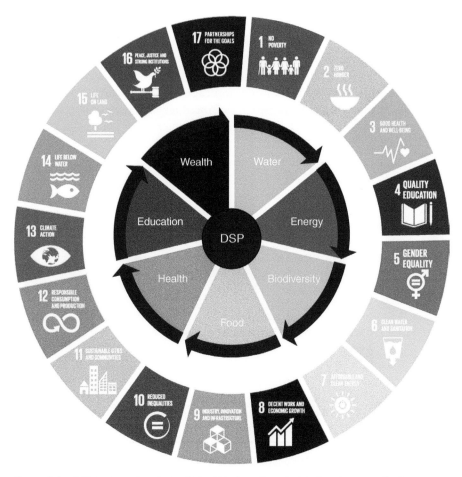

Figure 5.1   DSP is aligned with the United Nations Sustainable Development Goals.

investment in renewable energy not only addresses the goal of affordable and clean energy (SDG 7) but also contributes to climate action (SDG 13) and the promotion of sustainable cities and communities (SDG 11). This prevents isolated interventions and promotes a more balanced and sustainable strategy (van Tulder, 2018).

The SDG Compass was developed by GRI, the UN Global Compact and the World Business Council for Sustainable Development (WBCSD) to provide guidance for companies regarding how their strategies can be aligned with the SDGs (Global Reporting Initiative, UN Global Compact, World Business Council for Sustainable Development, 2015).

Ultimately, DSP applies SDGs to facilitate the prioritisation of business cases for potential investments with positive social and environmental impacts. The SDGs address pressing challenges such as poverty, inequality and climate change. By directing resources towards projects that align with the SDGs, we can channel them into areas where they are most needed and have the greatest potential for transformative change. This approach not only benefits society and the environment but also enhances the long term sustainability and resilience of investments.

How Does DSP Relate to the UN SDGs?

DSP recognises that resource rich regions should function as integrated economic, social and ecological systems. It considers key elements such as water, energy, biodiversity, food, health, education and wealth as initial indicators for sustainable development. This approach aims to enhance the prosperity of these regions, both during and after the extractive industry's operations. DSP looks beyond mere financial aspect of prosperity, delving into the regions' potential alignment with the UN SDGs, as shown in Table 5.1.

Table 5.1 How DSP addresses the United Nations Sustainable Development Goals.

DSP values	How DSP addresses United Nations Sustainable Development Goals
Water	**Reviving the Water Cycle:** How DSP addresses abundant and clean water for all, including conservation and restoration to support: 6 Clean Water and Sanitation 12 Responsible Consumption and Production 13 Climate Action 14 Life Below Water 17 Partnerships for the Goals
Energy	**Powering Sustainable Futures:** DSP's innovative solutions for abundant affordable and green energy to support: 7 Affordable and Clean Energy 11 Sustainable Cities and Communities 12 Responsible Consumption and Production 13 Climate Action 17 Partnerships for the Goals
Biodiversity	**Fostering Biodiversity and Ecosystem Harmony:** DSP employs Integrated Natural Resources Models to explore the relationships between subsurface and surface dynamics with climate influences to support: 11 Sustainable Cities and Communities 13 Climate Action 14 Life Below Water 15 Life on Land 17 Partnerships for the Goals
Food	**Nourishing the Future:** DSP contribution to SDGs through sustainable precision agriculture, enabling abundant and clean water, clean, abundant and affordable energy and sustainable value creation to support: 2 Zero Hunger 12 Responsible Consumption and Production 17 Partnerships for the Goals
Health	**Fostering a Future of Wellbeing:** DSP promotes quality of life for a future sustainable society, which supports: 3 Good Health and Wellbeing 11 Sustainable Cities and Communities 17 Partnerships for the Goals

Table 5.1   (Continued)

DSP values	How DSP addresses United Nations Sustainable Development Goals
Education	**Building a Strong Foundation for Future Generations:** DSP's dedication to enhancing education, cultivating skills and fostering a culture of lifelong learning, to support: 4 Quality Education 5 Gender Equality 9 Industry, Innovation and Infrastructure 10 Reduced Inequalities 13 Climate Action 16 Peace, Justice and Strong Institutions 17 Partnerships for the Goals
Wealth	**Fostering Sustainable Prosperity:** DSP's collaborative efforts to unlock the full human and natural resource potential, achieve holistic and sustainable wealth and ensure prosperity for all, which support: 1 No Poverty 7 Affordable and Clean Energy 8 Decent Work and Economic Growth 9 Industry, Innovation and Infrastructure 10 Reduced Inequalities 12 Responsible Consumption and Production 17 Partnerships for the Goals

The impact of investment in the extractive industry on the economic, social and ecological systems can be assessed through a blend of quantitative and qualitative metrics. The significance of these metrics varies depending on the region.

These measures facilitate an evaluation of the region's potential for future prosperity and the identification of areas for improvement. Recognising the varying significance of these aspects in different regions, it is imperative for key parties to be embedded and empowered from day one, actively contributing to the solutions tailored to the unique needs of each region.

Various methods can visualise a region's potential, with a spiderweb plot, as depicted in Figure 5.2, serving as an effective means to gauge its current standing against chosen metrics. The percentage reflects the current situation relative to the region's potential. This graphical representation highlights performance gaps, indicating opportunities for enhancement. To illustrate this approach, we present three case histories from three different copper regions.

- **Arequipa, Peru:** A vibrant, densely populated commercial and industrial centre, boasting the highest recorded levels of solar radiation in Peru and South America. Arequipa reported a population of approximately 970,000 people in 2024.
- **Skelleftea, Sweden:** An innovative, less densely populated industrial and mining city, home to prominent companies like Boliden AB (large mining and smelting company), Skellefteå Kraft (largest power company in Skellefteå) and Northvolt (lithium-ion battery cell manufacturing company). Skelleftea population reached around 76,000 people in 2024.
- **Smithers, Canada:** A town with a mining history in British Columbia, situated between Prince George and Prince Rupert. Surrounded by natural beauty, it falls within the Bulkley Valley, which is the traditional territory of the Wet'suwet'en. Smithers reported a population of about 5400 people in 2024.

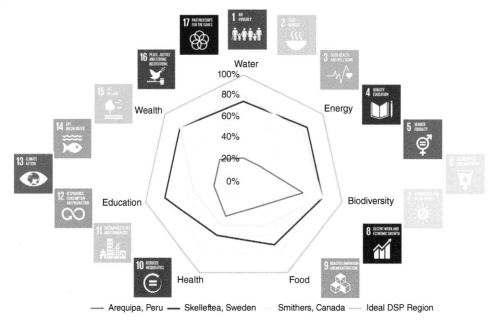

Figure 5.2 Assessment of three copper-producing regions' performance against the issues.

This preliminary assessment combines subjective and objective views based on measurable and unmeasurable parameters. The relative performance for these three regions is shown in Figure 5.2.

Based on this initial assessment, Arequipa emerges with the greatest potential for sustainable development. Skelleftea and Smithers, although ahead, still have the potential to develop resilience plans.

DSP's objective is to develop a prosperous region and secure a bright future for its inhabitants. The vitality of a region's community lies in its ability to retain talented individuals within the region, ensuring they do not feel compelled to seek better opportunities in major cities.

But, What About Financial Performance? How Does DSP Impact Return on Investment?

The process of prioritising investment decisions based on the SDGs can attract additional funding and support. Numerous investors, including impact investors, sustainable finance institutions and development banks, have incorporated the SDGs into their investment strategies. Demonstrating a commitment to the SDGs enables businesses and projects to tap into a broader pool of capital, attracting investors who prioritise sustainable development. This increased financial support can expedite the implementation of SDG-related initiatives, amplifying their impact, and ultimately contributing to the broader achievement of the global goals. DSP, therefore, brings added value and positively influences the return on investment (ROI).

The extractive industry has the potential to employ DSP principles ensuring the region maintains a viable and sustainable economy even after operations conclude. By serving as an enabler and a catalyst, the extractive industry can actively contribute to creating an environment that attracts other investors to participate in diversifying the regional economy. This multifaceted approach enhances the overall appeal of the region, drawing in additional investors and reducing risk for all. In this collaborative effort, the development of the region becomes a shared responsibility, less reliant on a single investor, thus minimising the likelihood of conflicts, while fostering a sustainable and diversified economy that benefits all.

References

Azapagic, A. (2004). Developing a framework for sustainable development indicators for the mining and minerals industry. *Journal of Cleaner Production*, 12(6):639–662.

Carvalho, F. (2017). Mining industry and sustainable development: Time for change. *Food and Energy Security*, 6(2):61–77.

Dehghani, H., Bascompta, M., Khajevandi, A. A., & Farnia, K. A. (2023). A mimic model approach for impact assessment of mining activities on Sustainable Development Indicators. *Sustainability*, 15(3):2688.

Global Reporting Initiative, UN Global Compact, World Business Council for Sustainable Development. (2015). *Learn more about the SDGs*. Retrieved from SDG Compass: https://sdgcompass.org/sdgs/

Han, X., Cao, T., & Yan X. (2021). Comprehensive evaluation of ecological environment quality of mining area based on sustainable development indicators: a case study of Yanzhou Mining in China. *Environment, Development and Sustainability*, 23(5):7581–7605.

United Nations Department of Economic and Social Affairs. (2015). *17 Sustainable Development Goals (SDGs)*. Retrieved from Sustainable Development: https://sdgs.un.org/goals

van Tulder, R. (2018). *Business and The Sustainable Development Goals: A Framework for Effective Corporate Involvement*. Rotterdam: Rotterdam School of Management, Erasmus University.

6

Nature's Hidden Gifts – Envisaging the Unseen Beneath Our Feet

Richard Blewett[1] and Doris Hiam-Galvez[2]

[1] *GeoSystems Consulting, Geoscience, Royalla, New South Wales, Australia*
[2] *Hatch Ltd, Designing Sustainable Prosperity, Vancouver, British Columbia, Canada*

Looking upwards into the sky, we can easily see the sun, the moon, the planets, the stars as well as the clouds and rainbows. Gazing skywards can generate feelings of a deep sense of space and trigger a wonder about the mysteries of the universe and how they relate to one's existence.

Contrast that with the feelings from looking down at the Earth beneath our feet. People may not feel the same curiosity or awe they do about the skyscape. They may see only dirt, rocks, grass and concrete. They may think that the Earth beneath their feet is dull and boring, compared to the sky above them. They may not realise that there is a whole world of geology hidden under the surface, with its own beauty, complexity, awe and possibility.

This lack of appreciation of the hidden subsurface is evident in the poor rendition of artificial intelligence-generated images of what is below us. Ask AI (DALL-E3 using Bing) to create a picture and the result is usually a visually attractive image (Figure 6.1) but unrealistic picture of what is down there. Part of the problem is the lack of realistic images for the AI database to train on, which reflects the general poor awareness of the true but hidden subsurface.

Holding a rock in one's hand can be a wondrous thing as every rock is like a book with its own fascinating story. A rock is like a book and a region with all its rocks and landscapes is like a library with many wondrous stories. Unlike a book or a library, the stories of the rocks are told through the minerals, textures, structures and fossils – in other words, the language of Geology.

Figure 6.1 AI-generated image using the request 'Can you generate a realistic image of the subsurface of the Earth please. Do not have labels'. The image is attractive but not realistic of the subsurface. *Source:* Bing (2023).

What is Geology?

Geology is the science that studies the structure, composition, history and processes of the Earth. It reveals how the Earth was formed, how it changed over time and how it continues to change today. It is the grandest of stories told to us through the layers of rocks that record the events of millions and billions of years ago, such as volcanic eruptions, earthquakes, mountain building, erosion and sedimentation.

Geology provides us with the minerals and much of the energy we use. It also provides us with the fossils that tell us about the diversity and evolution of life and environments that existed in different periods of Earth's history. Geology tells us about us and where we come from and why.

Geology shows us the movements of tectonic plates that shaped the continents and the oceans, which drove the climate now and in the past. Such climate changes drove the evolution of life, including our own species out of Africa to settle and occupy the entire planet. Geology also provides us with the essential cycles of water, carbon, nitrogen and other elements that sustain life on Earth.

Geology is not only a scientific discipline; it is also a part of art and philosophy. It helps us appreciate the beauty and diversity of nature and understand our place in it. It helps us respect and protect the Earth and its resources. It helps us imagine what lies beyond our sight and what may lie in the future.

The next time you look down at the ground, try to see beyond the dirt. Try to imagine what is below you in terms of geology. You may discover a new perspective on yourself and your world.

A Shaping Force

The rocks beneath our feet are not static, but constantly moving and changing due to plate tectonics, erosion, weathering, folding, faulting, volcanism and other processes. These processes create various geological and landscape features, such as mountains, valleys, rivers and volcanoes. The weathering and movement of rock and soil create the substrate for our food. The shape of the land and its position on the globe dictate how water, life's essential ingredient, is stored and transported either as rivers or hidden beneath the surface as groundwater. Geology has been a shaping force in life's evolution, including our own. It has created the world that enabled the development and sustenance of the great civilisations, including the present.

The subsurface also contains many natural resources that are essential for human civilisation, such as minerals, oil and gas, coal and groundwater. Minerals are solid substances that have a specific chemical composition and crystal structure. They can be extracted from rocks or ore deposits by mining or quarrying. Oil and gas are liquid or gaseous hydrocarbons that accumulate in porous rocks or reservoirs. They can be extracted by drilling wells and pumping the gas/oil to the surface. Coal is a solid fossil fuel that forms from the burial and compression of plant matter. It can be mined by surface or underground methods. Groundwater is water that fills the spaces between soil and rock particles or rock fractures. It can be accessed by digging wells or using pumps from bores to bring water to the surface.

Seeing the Unseen

We cannot see directly what lies beneath the soil and rocks, but we can use various methods to infer and visualise the subsurface structures and properties. These techniques are analogous to medical imaging like X-rays or CT scanning of the body. The remote sensing technologies that probe and map the Earth use the natural properties of the rocks and minerals such as density, magnetism, radioactivity and electrical conductivity to create beautiful 3D maps and images that are wondrous to see.

We can image small areas at very high-resolution, or vast volumes including continents (Figure 6.2) or even the entire Earth. Techniques can be applied either on the ground or from the air using aircraft or drones, or from space using satellites. They include, but are not limited to:

- **Seismic Surveys:** These use sound waves to create images of the subsurface layers and structures. The sound waves are generated by explosives or vibrators on the surface or in boreholes and are recorded by sensors called geophones or hydrophones. The sound waves reflect or refract when they encounter different rock densities and boundaries between rocks of different densities. By analysing the travel time and amplitude of the sound waves, we can infer the depth, thickness, shape and composition of the subsurface layers and structures.
- **Gravity Surveys:** These measure the variations in the gravitational field of the Earth caused by differences in densities of the subsurface materials. The gravity field is

measured by instruments called gravimeters on the surface or in aircrafts or satellites. The gravity field is affected by factors such as elevation, latitude, topography and tides. By correcting for these factors, we can isolate the gravity anomalies caused by subsurface features, such as faults, basins, different rock types or even ore deposits and water.

- **Radiometric Surveys:** These use the natural decay of radioactive isotopes present in all rocks. The main elements measured include potassium, uranium and thorium, which vary according to primary rock chemical composition and the weathering of these rocks. Wonderfully coloured maps of the near surface are generated by combining these measured concentrations into a single image.

- **Magnetic Surveys:** These measure the variations in the magnetic field of the Earth caused by differences in magnetisation of the subsurface materials. The magnetic field is measured by instruments called magnetometers on the surface or in aircrafts or satellites. The magnetic field is affected by factors such as latitude, declination, inclination and diurnal variations. By correcting for these factors, we can isolate the magnetic anomalies caused by subsurface features, such as igneous rocks, iron ores or magnetic minerals.

- **Electrical Surveys:** These measure the variations in the electrical conductivity or resistivity of the subsurface materials. The electrical conductivity is measured by electrodes on the surface or in boreholes, or sensors in the air on aircraft. The electrodes generate a current into the ground and measure the effect of the electrical flow (or not) through the Earth. The electrical conductivity is affected by factors such as moisture content, salinity, temperature and porosity of the subsurface materials. By analysing the distribution and magnitude of the electrical conductivity, we can infer the presence and location of subsurface features, such as groundwater, clay layers, salt domes or metallic ores.

- **Radar Surveys:** These use radio waves to create detailed images of the subsurface features and properties, usually over small areas. The radio waves are generated by transmitters on the surface or in boreholes and are recorded by receivers on the surface or in boreholes. The radio waves reflect or scatter when they encounter different rock types or boundaries. By analysing the travel time and amplitude of the radio waves, we can infer the depth, shape and composition of the subsurface features.

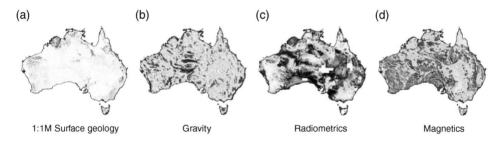

(a)   (b)   (c)   (d)

1:1M Surface geology   Gravity   Radiometrics   Magnetics

Figure 6.2 Different views of the Geology of Australia surface geology (a), gravity (b), radiometrics (c) and magnetics (d). *Source:* Geoscience Australia (2018) / CC BY 4.0.

None of the above techniques tells us what the hidden rocks and materials are. As with medical imaging and diagnostics, we need a biopsy or a sample to test further in the laboratory to truly understand what we cannot see. In geology, we drill boreholes to obtain samples for further analysis, which are used to validate and test the maps and images.

The depth of the drill hole varies depending on the science question or target. The deepest research drill holes have only penetrated 12 km into the Earth, which is only a third of way to the base of typical continental crust. Most drill holes in oil and gas fields are around 1–3 km deep and in mining areas less than 1 km (Figure 6.3). Drilling for groundwater is typically restricted to only a few hundred metres. Given we have but mere 'scratching' into the Earth, there is still a vast amount of the subsurface that we know only through remotely sensed images.

Figure 6.3   Drilling for mineral exploration in outback Australia. Note that no rocks are visible at the surface and so to 'see' what is below requires remote sensing followed by drilling. *Source:* Adwo/Adobe Stock Photos.

How Does Seeing the Unseen Deep Geology Help with DSP?

Most mining regions are known for one or few commodities, be they copper or gold or uranium or the emerging critical minerals. This means that mining companies become very good at focussing on discovery and recovery of these limited commodities in any region, irrespective of the potential for more diverse and sustained development.

Having a holistic understanding of the broadest resource potential can, and should be, generated for each region. Having an open mind shaped by a broad understanding of what is possible enables communities, investors, developers, governments and regulators to make the best-informed decisions for the sustainable prosperity of the region. Consideration of possibilities should include the full gamut of mineral, energy, storage, water and infrastructure development diversity.

Mineral Diversity

Despite the dominance of a few commodities within any region, the span of commodity options available can be far greater than first envisaged by those developing it. For example, the Eastern Goldfields region of southwest Western Australia (see case study) is well known for its gold and nickel. Many of the gold deposits lie hidden beneath the soil and desert cover materials that include salt lakes, which were largely ignored as having any resource value. Analysis of salty brine waters in these lakes shows that they are particularly enriched in potash (potassium), which is an essential ingredient in fertiliser and was previously imported into Australia. This previously unknown opportunity has now transformed into value for not only the company but also the local people who have jobs and opportunities (Australian Potash, 2023).

The global energy transition is driving a change in the increasing demand for materials previously considered curiosities of the Periodic Table. These commodities include the Rare-Earth elements and other so-called critical commodities that are needed for defence and space industries, as well as electric vehicles and renewable energy power generation and transmission. These commodities are commonly found as companion elements in the traditional ore bodies with the former being ignored and discarded into waste and tailing dumps.

There is an important role here for geologists and governments to ensure a full inventory of the commodity potential for a region is established.

Energy Diversity

The subsurface also is a source of new forms of energy. These can be from naturally occurring hydrogen gas or from the natural heat emanating out of the Earth.

Hydrogen is known in a range of differently named colours. We have green hydrogen, which is made by splitting water into oxygen and hydrogen; blue hydrogen is extracted from natural gas with the associated carbon emissions captured and stored and there is grey hydrogen which also uses natural gas but where the resultant CO_2 is released to the atmosphere.

There is also white or gold hydrogen, which is produced deep in the Earth through degassing from stores in the crust and mantle and by hydrolysis of water on its interaction with specific types of rocks at shallower levels. Hydrogen is being discovered in locations where it was previously unknown or dismissed as a nuisance gas. Commercial quantities of hydrogen were discovered by accident while drilling for water in Mali in the 1980s. Hydrogen can be found in old or unviable oil and gas provinces and even in regions better known for its nickel and iron ore mines.

Another form of energy from the deep Earth is geothermal. Depending on the geological region, geothermal heat is generated by either magmatism or by the natural decay of radioactive minerals. Regions with active volcanoes such as Iceland, New Zealand, the United States, the Philippines and Mexico have geothermal power stations that in 2023 have generated a massive 95 billion kWh of electricity (Statista, 2023).

In areas of older crust where present day volcanism and magmatism are not present, the natural geothermal gradient where heat increases with depth in the Earth can provide a source of energy. The natural decay of radioactive elements in the crust can, depending on the specific geology, be enough to create hot water or hot rocks at depths shallow enough to generate power. In some places, these hot waters can come to the surface and be expressed as hot springs, which can be a tourist and health destination.

Knowing and measuring the geothermal energy potential of a region could provide noncarbon opportunities for baseload power generation or supplementation.

Helium

Although not an energy source, helium can also be present in oil and gas fields. It is a critical element in some national strategic minerals lists. Helium is used in party balloons, but it has serious use in analytical instruments, cryogenics, MRI scanners, quantum computing and particle accelerators.

Helium can be measured remotely and predicted based on the known geology of a region, thus providing another potential source of industry diversity for a region.

Storage

We commonly think of the extraction of materials and commodities from the Earth. However, we also can use a region's geology for emplacement or storage of materials on a temporary basis or a more permanent one. The type of subsurface will dictate what is possible, but fortunately, there are many potential options.

The drive for net zero is opening regions for the permanent sequestration or storage of CO_2 in sedimentary basins beneath the Earth. The CO_2 being stored can be from an industrial point source like a traditional power station, steel works, oil/gas field, or cement factory, or it could be directly from the atmosphere as direct-air capture (DAC) systems. Pipelines and pumping infrastructure are required to transport the point source-generated CO_2 to a favourable geological site of capture. They are commonly not adjacent to one another.

In contrast, the advantage of DAC systems is that they can in theory be deployed anywhere directly above geology that is sufficiently favourable for storage. This is especially so where the solar/wind resources are abundant and used to power the operation of DAC systems themselves.

In all instances, the storage of gas like CO_2 underground requires geological conditions that are found in many but not all sedimentary basins. There needs to be sufficient space between the particles making up the rock – called porosity and there needs to be sufficient connectivity between the pores – called permeability. There needs to be sufficient porosity and permeability – called a reservoir – at a suitable depth (>1 km) to keep and prevent the buoyancy of CO_2 from rising due to gravity. The reservoir also needs to be capped by seal-like rock – usually clays or shales – that are impermeable to the flow of gas. The area also needs to be free of disruption by faults that 'break' the seal and could cause gas leakage.

In some sedimentary basins, salt has accumulated from ancient oceans into thick layers now buried beneath the surface. A number of these salt accumulations have been mined for consumption and industrial uses. Salt not only makes food taste great, but it also makes a good storage medium. Salt can be used as a permanent store of highly hazardous wastes including nuclear materials. Salt can also be used for temporary storage of gas, such as hydrogen or helium (Figure 6.4). Regions with salt in them have many opportunities for different industries, whether that be in present day lakes such as table salt, lithium, boron, uranium or potash, or when buried as a safe storage medium.

The permanent storage of waste and hazardous materials is an ongoing challenge for society. A region may not have the salt layers or the porosity of a basin to store gas like CO_2; it may be one of the ancient cratonic or shield areas that form the backbone of most continents. These regions are generally tectonically stable – being distant from the geohazards of earthquakes, volcanoes and tsunamis – and have been for many millions or even billions of years. These regions can be investigated for storage of particularly hazardous materials, such as high-grade radioactive waste.

Figure 6.4 Conceptual diagram of a salt diapir suitable for development for storage of hydrogen or other materials. *Source:* Commonwealth of Australia (Geoscience Australia) / Public domain / https://www.ga.gov.au/scientific-topics/energy/resources/hydrogen/australias-hydrogen-production-potential.

Groundwater in the Water Cycle

Water is essential for life. So, the responsible use and management of water are essential for any development. Mark Twain is reputed to have once said that 'whiskey is for drinking and water is for fighting'. Conflict around development can be linked to several issues, but a region's ongoing water security is a consistent theme where communities, regulators and developers can agree.

In many parts of the world, there are severe demands on the available water. Many areas are literally mining their water to such an extent that the land surface is falling due to water extraction at alarming rates. In these regions, the aquifers holding the water are being depleted at rates greater than their recharge. Going forward, we are likely to see the situation worsen as temperatures rise due to global warming, leading to enhanced evaporation and depending on the region, increased droughts (Figure 6.5).

We can all see water falling from the sky as rain or snow and gathering into lakes and rivers that flow downslope into the sea, or in rare places into inland seas and basins. What we do not usually think about is the water that lands on the ground and is absorbed to become the groundwater. Surface water and groundwater are tightly coupled and are key ingredients in the water cycle (Figure 6.6).

Water Stress by Country: 2040

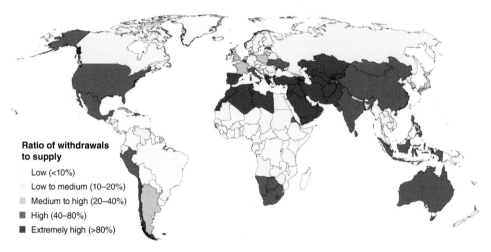

NOTE: Projections are based on a business-as-usual scenario using SSP2 and RCP8.5.

Figure 6.5   Prediction of regions of water stress in 2040. *Source:* World Resources Institute (2015) / with permission of World Resources Institute.

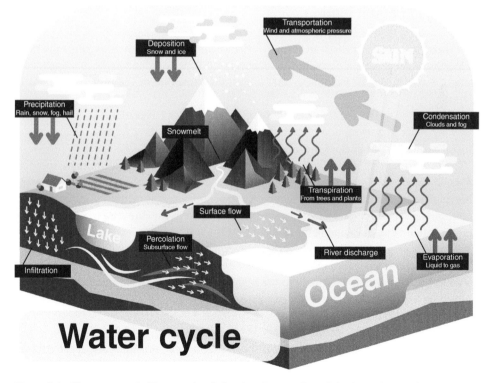

Figure 6.6   The water cycle illustrated and showing the coupling of the flow of water from the oceans to the land via the atmosphere and its return via surface run off or groundwater flow. *Source:* Adobe Stock.

Groundwater is one of the most important resources that we obtain from the geology beneath our feet. Groundwater provides drinking water for millions of people around the world, as well as irrigation water for agriculture and industry. Groundwater also sustains ecosystems such as wetlands, rivers and lakes and maintains the base flow of streams and springs. Groundwater can be a renewable resource, but it is also vulnerable to overexploitation, contamination and depletion. Therefore, we need to manage and protect groundwater wisely and sustainably.

We can use many of the same techniques that remotely image the Earth to map and understand a region's groundwater. Groundwater does not lie in underground lakes, nor does it flow like we understand it does in rivers. The water percolates into the little spaces between the grains of the rocks and soil, or into fractures and joints in impermeable rocks like granite. This means that groundwater is unevenly distributed in the subsurface and such distribution in terms of the aquifers that contain the water and the aquicludes that form barriers to water flow must be mapped and understood.

As with the storage of gas, water can be actively engineered to be stored in aquifers where the geology permits. We call this managed-aquifer recharge (MAR) where excess surface water – maybe generated during periodic flooding – is not wasted but is directed to recharge areas of an aquifer for underground storage to be later used as needed. Understanding the geological viability of MAR could be the answer for some regions in securing their water.

New Ways of Metal Mining

There is also an opportunity to transform the way we mine today by extracting only the value and leaving the host rock in place. In situ recovery or ISR, as described in the metals chapter, is one of the advances in technology that is set to become one of the mining methods of the future.

As the name suggests, ISR involves the selective removal of target elements and minerals by drilling, rock fracturing and chemical leaching directly at the drill site, which creates minimal surface disturbance of the land surface.

Effective fragmentation and containment are pivotal in ISR to optimise mineral recovery and avoid contamination (Hiam-Galvez, Harding, Krawchuk, Slabbert, & Brown, 2018). Achieving this entails a deep understanding of the ore body and surrounding geology. When natural geological formations like low permeability strata envelop the ore body, they provide natural containment, allowing better control over the leaching process. If the permeability of the surrounding rock is lower than the mineralised rock, the orebody is naturally contained which allows injection and recovery well designs that can provide good hydrological control as fluid movement is limited to the higher permeability areas. In scenarios lacking such natural barriers, innovative engineering solutions come into play (Hiam-Galvez, Gerber, Pekrul, & Earley, 2020).

The success of ISR lies in the synergy between geoscientists and process engineers. Together, they unlock the full potential of the subsurface, determining the most efficient extraction methods. This collaboration is like the precision of the hummingbird extracting nectar illustrated in the front cover of this book, where every action has a purpose and contributes positively to the environment.

Knowing the Unknown – An Aid for Infrastructure Planning

The planning and construction of any infrastructure require knowledge of the subsurface. This is why civil engineers are increasingly creating digital twins of the Earth to facilitate infrastructure planning, identify potential issues with challenging ground conditions and adjust the design and/or route of the infrastructure to minimise delays and excessive costs (Liu, Zhang, & Xu, 2023).

Similarly, knowing the potential geohazards from the subsurface is crucial for the siting of big infrastructure such as dams, powerplants and waste repositories. Knowing the overall earthquake risk, together with the presence of likely causative structures (faults), is a critical piece of information to have for any planning.

The expansion of renewable energy projects has meant that new and additional transmission lines are needed to supply renewable power from the more widely distributed solar, wind, tide or geothermal energy generation sites to the demand centres. Where to place the transmission lines is not just a matter of working around the landscape and communities impacted but also what is beneath the route. The Earth has natural resistors and conductors within, which impact the flow of electricity in transmission lines when certain geomagnetic (solar) storms occur (Cordell, et al., 2021). Knowing the geoelectric heterogeneity of the Earth can avoid serious power interruptions or failures and blackouts.

Final Words

Will Durant, the US philosopher once said that 'civilisation exists by geological consent: subject to change without notice'. Geology matters to us all. So, the next time you look down, think not of dirt, rocks and soil but of a place of wonder and possibility just waiting for its story to be told.

References

Australian Potash. (2023). https://www.australianpotash.com.au/site/content/

Bing. (2023). *DALL-E* (Version 3). https://www.bing.com/images/create

Cordell, D., Unsworth, M., Lee, B., Hanneson, C., Milling, D., & Mann, I. (2021). Estimating the geoelectric field and electric power transmission line voltage during a geomagnetic storm in Alberta, Canada using measured magnetotelluric impedance data: The influence of three-dimensional electrical structures in the lithosphere. *Space Weather*, 19(10):e2021SW002803.

Geoscience Australia. (2018, September). *Exploring for the future: Realising the resource potential of the NW mineral province (and beyond)*. Retrieved from University of Queensland: https://smi.uq.edu.au/files/25158/09_NW%20Minerals%20Province%20talk%20Blewett%202024%20Sep%202018.pdf

Geoscience Australia. (2023, October 24). *Australia's hydrogen production potential*. Retrieved from https://www.ga.gov.au/scientific-topics/energy/resources/hydrogen/australias-hydrogen-production-potential

Hiam-Galvez, D., Gerber, E., Pekrul, J., & Earley, D. (2020). In Situ Recovery (ISR) – The Permitting Challenge. *ALTA 2020 In-Situ Recovery Conference* (pp. 64–73). Melbourne: ALTA Metallurgical Services Publications.

Hiam-Galvez, D., Harding, D., Krawchuk, P., Slabbert, W., & Brown, R. (2018). An Alternative Mining Concept. Vancouver: Canadian Institution of Mining, Metallurgy and Petroleum.

Liu, C., Zhang, P., & Xu, X. (2023). Literature review of digital twin technologies for civil infrastructure. *Journal of Infrastructure Intelligence and Resilience*, 2(3):100050.

Statista. (2023). *Geothermal Energy – Worldwide*. Retrieved from https://www.statista.com/outlook/io/energy/renewable-energy/geothermal-energy/worldwide

World Resources Institute. (2015, August 26). *Ranking the World's Most Water-Stressed Countries in 2040*. Retrieved from https://www.wri.org/insights/ranking-worlds-most-water-stressed-countries-2040

7

The Method – Designing Sustainable Prosperity

Doris Hiam-Galvez

Hatch Ltd, Designing Sustainable Prosperity, Vancouver, British Columbia, Canada

Introduction

Essence of the Method

In this chapter, we delve into the heart of DSP (Hiam-Galvez, Prescott, & Hiam, 2020), uncovering its systematic approach to change the thinking process and foster the creation of sustainable regions. We embark on a structured three-phase planning process: building the design inputs by gathering data, designing system solutions and preparing the environment for efficient implementation.

At the core of DSP is a dynamic seven step design process. We begin by preparing teams for collaboration and nurturing a visionary outlook for the region. We then analyse data, identify market needs and unlock regional potential. Crafting solutions and action plans follows, alongside adapting the education system for future readiness. Going beyond planning, we create an environment to accelerate execution, foster innovation and actively engage key parties. Our aim? Sustainable development, maintaining momentum and seizing new opportunities for positive regional impact.

Mining as a Catalyst for Economic Diversification

In resource rich regions, the local economy relies on resource extraction as its primary driver. However, as illustrated by the red curve in Figure 7.1, there is an inherent limitation to this model: these resources are finite in nature. Whether their life span stretches over half a century or even more, their end is inevitable. The red curve captures this trajectory, highlighting how a thriving resource extraction activity amplifies local economic activity. But as the mine reaches the end of its life, so does the local economic activity.

Supporting the extractive industry, we have ancillary industries like suppliers and service providers, represented by the yellow curve. They, too, follow the rise and eventual fall of the resource extraction activity.

However, DSP introduces the green curve. This represents the future pillars of sustainable prosperity – one that, though sparked by extractive industry, is not dependent on it. Here, the

Designing Sustainable Prosperity: Natural Resource Management for Resilient Regions, First Edition. Edited by Doris Hiam-Galvez.

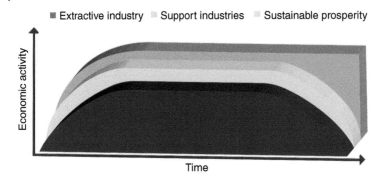

Figure 7.1 Designing Sustainable Prosperity (DSP).

extractive industry acts as a catalyst, with one of the primary objectives being to lay the foundation for a thriving sustainable economy.

DSP offers a paradigm shift, urging regions to look beyond extractive-focused prosperity and embrace a broader, more inclusive regional prosperity vision, depicted in Figure 7.2.

By reshaping regional perspectives and fostering collaborative efforts, DSP has the potential to guide resource rich regions towards a sustainable and prosperous future. The region transitions from a potentially finite life economy to a long-lasting prosperity.

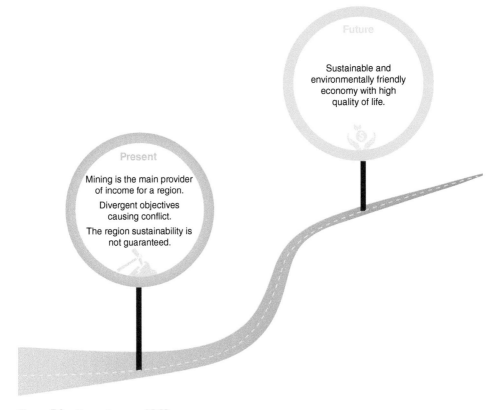

Figure 7.2 The outcome of DSP.

Significance and Impact of the DSP Approach

The DSP approach stands out as a new way to achieve sustainable development. Its significance and impact are rooted in its core elements, each contributing to a holistic, long term vision for regional prosperity. Here are these core elements:

- **Vision and Strategy**
  - Long term Vision: Focusing on regional prosperity. This approach minimises conflict, promoting a unified effort towards common goals.
  - Economic Diversification: Transitioning to a knowledge-based economy for sustainable long term prosperity.
  - Systems Design: Treating regions as interconnected ecosystems to ensure holistic, synergistic solutions.
- **Method Approach**
  - Multidisciplinary System: Integrating technical, social, environmental, financial and political dimensions to unearth hidden regional potentials for sustainable solutions.
  - Customisation for Regional Needs: Tailoring solutions aligned with market needs.
- **Collaboration and Empowerment**
  - Collaborative Transformation: Co-creating a shared vision and actionable strategies for success.
  - Community Empowerment: Encourages collaborative transformation and community empowerment, underpinned by a fresh perspective that is essential to create the conditions for a sustainable region.
- **Focus on Key Resources and Alignment with Global Goals**
  - Focus on Essential Resources: Where appropriate, employ clean energy and water as the foundation to enable other sustainable enterprises.
  - Nature Positive Solutions: Developing enterprises that contribute to ecological regeneration.
  - Alignment with Global Goals: Adhering to the UN SDGs as metrics for economic, social and environmental prioritisation.

The significance of the DSP approach lies in its ability to transform regions through a structured, collaborative and holistic method. By integrating these core elements, DSP creates a guide for key parties to navigate the complexities of sustainable development. This section delves into each element, exploring its impact and how it contributes to the overall effectiveness of the DSP approach in fostering sustainable regional prosperity.

Incorporating these core elements into the DSP approach also addresses specific challenges in rural regional contexts. For instance, a common issue in remote resource rich regions is the reliance on fly-in and fly-out (FIFO) workforce models. This practice often results in the workforce being located outside the region, creating a disconnect between the mining activities and the local community. Anticipating future trends, DSP highlights the potential increase of remote operating centres located outside the region. This shift necessitates tailored solutions to address the unique challenges of each region. For example, one case study illustrates establishing remote operating centres closer to or within these remote mining locations. Such initiatives aim to bridge the gap between these activities and the local community, fostering a deeper connection and understanding.

The DSP process encourages a paradigm shift in perspective, urging key parties to think creatively about the possibilities within a region's human and natural resources, whether it's leveraging local talent, utilising unique geological features or capitalising on regional strengths.

Importantly, DSP is a versatile and region independent method. This flexibility allows the process to be applied in various contexts, across different regions, making it a universally relevant approach to sustainable development.

The collaborative spirit underpinning DSP fundamentally redefines the relationships between key parties, fostering trust and aligning mutual objectives. Regions embracing the DSP approach are likely to witness significant uplifts in community well-being, ecological balance and economic resilience.

Ultimately, the DSP method empowers regions to not just navigate but thrive in a future beyond resource extraction. By weaving together, the social, environmental and economic threads, DSP crafts a pathway for sustainable prosperity, ensuring regions are equipped to thrive long after resource extraction activities have ceased.

Pioneering the Knowledge Economy

One potential outcome of the DSP process is the development of a knowledge economy. The emphasis is on nurturing local knowledge that is conceived, honed and retained within the region and then showcased to the global marketplace. One objective is to foster specialised knowledge that paves the way for resilient and sustainable enterprises. It also provides focus to compete on the global market, offering high value.

Planning Process

Leveraging the expertise of a multidisciplinary team, including local key parties, this process creates a pathway for a resilient and sustainable region where the extractive industry is one of the components of the system. Key components of the process are:

- **Strong Leadership:** Effective leadership is crucial in guiding this transformative journey. The initial champion sets the direction, while the process itself cultivates additional leaders, reinforcing the collective drive for meaningful change and sustainable impact.
- **Integrated Approach:** A multidisciplinary team, including local key parties, is fundamental to the transformation.
- **Preliminary Foundation:** The initial months are reserved for creating a foundational baseline. This will serve as the cornerstone for subsequent designs and a reservoir of data for continually innovating future solutions. The focus is on understanding the data to visualise the hidden potential rather than just generating data.
- **Solutions System:** Following the baseline creation, the focus shifts to designing system solutions via the structured seven step process. The first two steps aim to empower the team to envisage an ambitious trajectory for both them and the region. Only when the team is open to collaboration and feels empowered can they truly co-create the future. After generating a plethora of ideas, typically, it's after exhausting initial thoughts that the most innovative and impactful ideas emerge.
- **Expanding the Awareness of Possibilities:** The approach progresses systematically through seven steps, each expanding the awareness of possibilities and fostering trust and collaboration, culminating in higher innovation levels.
- **Simultaneous Engineering:** Emphasising efficiency, simultaneous engineering (all phases of the project are considered throughout the process) ensures all key participants from all disciplines collaborate from start to finish, leading to holistic solutions and fostering ownership, which in turn ensures practical implementation. To eliminate the need to

rework, the models are built up to a certain level of completion and they continue being optimised as the design process progresses.

Phases of the Planning Process

The planning process commences with a champion, an individual deeply committed to the region's future, possessing both vision and determination to drive change. Champions can emerge from various key parties, including private investors, government officials, academics and community leaders. As the process unfolds, additional champions will emerge.

Funding for the planning phase can be secured from diverse sources, including:

1) **Private Investors:** These may include key players from the extractive industry including end users, green energy investors, major infrastructure developers or others with a keen interest in the region's long term success and a desire to mitigate risks.
2) **Philanthropy and Academia:** Initially, philanthropic seed funding and partnerships with university business innovation incubators can provide crucial support. As the planning progresses and economic viability is demonstrated, opportunities for venture capital, private equity and commercial banking involvement may arise.
3) **Public Sources:**
   a) Regional Government Funding: Regional governments can offer support in the early stages, aligning their efforts with regional industrial policies and the UN SDGs.
   b) National Governments: Governments seeking sustainable re-industrialisation and energy transition preparation may contribute, especially in areas like critical minerals extraction and economic diversification.
4) **Development Banks and Development Agencies:** Collaboration with development banks and agencies can provide access to additional funding resources.
5) **Innovative Financial Models:** Combining various funding sources or exploring innovative financial models can be considered to ensure the successful execution of the planning process.

The planning process spans many months. An initial vision can emerge within a few months; however, to explore the full spectrum of possibilities and craft an actionable pathway, it is reasonable to allocate from 9 to 12 months. Allowing sufficient time in the process is crucial for unleashing the full potential of both the individuals within the team and the region. This intentional space fosters creativity, encourages innovative thinking and enables the team to harness their collective capabilities, leading to a vision that embraces the possibilities for the region to compete in the world market.

As part of the planning process, the team should be assembled, comprised of champions who have a vested interest in the future of the region and understand what it takes to lead change, but also individuals from varying backgrounds are needed to unearth and harness the region's full potential. To capture the region's perspectives, it may involve key players such as:

- Community leaders
- Indigenous representatives
- Government officials
- Investors
- Academia
- Industry representatives
- Environmental groups
- Knowledge experts

The planning process encompasses the following phases, each consisting of several steps, summarised in Table 7.1 and described below:

Table 7.1 Summary of the planning process.

Planning process	Activities
Design Inputs: Socio-economic baseline	Establish a baseline of the current socio-economic situation: 1) Map a socio-economic baseline which will support a market study and provide insights for a preliminary understanding of the potential for the region. 2) Understanding the Community: Gather personal stories from residents to inspire a formulation of a bold vision for the region and society. 3) Understanding of the cultural values of the region.
Design Inputs: Potential mapping	Establish the design basis for the region: 1) Conduct a market study to identify the needs of the market as it relates to the potential of the region. 2) Determine the potential of the region by creating an Integrated Natural Resource Model (INRM) that includes sub-surface, surface and climate potential.
Design: System solutions	Determine the potential enterprises (products, technologies and services) that will become the pillars of the future economy following the seven steps: 1) Ignite individual and team potential. 2) Formulate a bold vision for the region. 3) Seeing data aligned with their vision, the team experiences a breakthrough, recognising their potential. 4) Design the Future: With an expanded awareness of possibilities, the team creates system solutions ideally suited to the region. Develops strategic business cases to build a resilient region. 5) Path to the Future: After prioritising the business cases, strategic path is defined for the region's sustainable expansion. 6) Relevant Education: Adapting the local education system to provide the skills and tools to support the future economy. 7) Action and Maintenance: Implementation plan including short-, medium and long term actions. Actions to create an environment to maintain the momentum.
Path of Impact:	Develop a roadmap by prioritisation of business cases using an integrated approach using the UN SDG indicators to include economic, technical, environmental and social impact. Select the preferred pathway for maximum positive impact.
Implementation Readiness:	Actions, to scale solutions for implementation, policy adaptations and education. Create an empowering environment where individuals can execute the plan, consistently delivering exceptional value to their target markets by harnessing the region's unique strengths.

1) **Building the Design Inputs**

   This phase is dedicated to gathering essential data from diverse sources and integrating them into simple and visually compelling formats. Through this process, we unlock the power of data where knowledge is key to make sense of the data. Transforming the data into a clear and understandable narrative sheds light on the region's potential and sparks fresh insights for the future of the region. These might include:
   1) Market study including the unmet needs of the net zero world and nature positive solutions
   2) INRM to determine the potential of the region

3) Socio-economic and environmental baseline
4) Capacity to transition to a green future
5) Cultural strengths and ancestral knowledge

The team engages in collecting the design inputs.

2) **Designing System Solutions**

Once the design inputs have been built and essential insights have been gathered, the planning process can proceed with a solid foundation.

The design phase includes a structured workshop following the defined steps to gradually 'expand the awareness of possibilities' to co-create a sustainable and resilient system for the region where the extractive industry is a catalyst. This includes prioritising the business cases consistent with the UN SDGs and selecting the path for the highest impact.

Unleashing the potential of both individuals and the region is the starting point for the designing process. The diverse team that has been gathered will have its own drivers and possibly opposing targets. Therefore, preparation is needed for the individuals to be open and willing to collaborate. Before entering the design stage, the team must be primed to welcome transformation.

- **Preparing the Team for Change (Steps 1 and 2):** These steps serve to prepare the team for effective collaboration. The first step focuses on unleashing the potential of each participant and the team, while step 2 leverages the knowledge gained in the first step to expand possibilities for the region. These preparatory steps set the stage for the collaborative design phase.
- **Design (Steps 3–7):** These steps represent the pivotal phase where the team designs innovative but practical solutions tailored to the region's needs, ensuring the development of sustainable systems. High level business cases are developed and ranked, attractive packages to attract investment are developed and the path for maximum impact is selected.

3) **Implementation Path:** Prioritise the business cases aligned with the UN SDG indicators to identify and select the most impactful pathways for scaling solutions for implementation.

4) **Implementation Readiness:** Creates the environment to maintain the momentum and start engaging with potential investors to implement the most impactful pathways.

Key Design Inputs

There are three primary inputs:

- An integrated natural resources model unveiling the region's potential.
- The socio-economic, environmental and cultural inputs (core values of the region)
- A market study identifying forthcoming needs.

Integrated Natural Resources Model

By delving into the geological history, the Integrated Natural Resource Model (INRM) reveals layered perspectives that uncover hidden potentials, including sub-surface, surface and climate impact. By bringing the unseen to the forefront, new sources of water can be uncovered, new energy sources and even assessing soil quality. A comprehensive insight, like an X-ray of the region's latent potential, harnessing the power of geoscience, earth science and various other scientific disciplines. Yet, understanding the data is more pivotal than merely generating it. Even in a world saturated with technology, insightful interpretation can be achieved with paper and pen in the hands of discipline experts combined with Indigenous knowledge to grasp the broader picture.

The INRM enables a comprehensive understanding of the potential available in a region. To construct an INRM, the information incorporated can vary based on the region's natural resource potential and characteristics. While data points are listed below, the approach is not exclusive and will adapt to the specific potential of each region. The INRM is therefore an amalgamation of diverse data sources, offering a comprehensive view of a region's potential. It may integrate the following data sources:

- **Earth data**, encompassing both:
  - Surface such as biodiversity, land use, surface water, solar radiation, soil quality.
  - Sub-surface such as mineral systems, sub-surface water, seismic activity, energy sources, storage potential.
- **Societal data metrics,** which cover:
  - Population metrics such as density, skillset, education, health.
  - Infrastructure details such as transport, energy production and distribution, centres of education, communication links.
  - Commercial factors such as political stability, conflict, investment attractiveness.
- **Climatic data**
  - Both current and future climatic factors such as wind speed, rainfall and pattern, climate change predictions, El Niño and La Niña, other macro- and micro-seasonal and climatic events.

For example, Figure 7.3 of the interconnected water system illustrates how INRM can provide a holistic understanding.

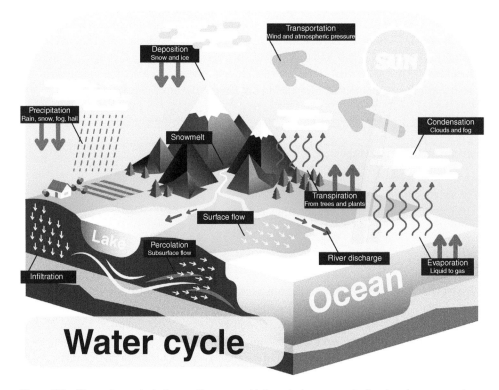

Figure 7.3   The water cycle is the continuous and integrated movement of water above, on and below the earth's surface. *Source:* VectorMine/Adobe Stock Photos.

Groundwater plays a pivotal role in moulding the surface landscape through various weather-related and geomorphological activities. Our rivers, lakes and wetlands are in a perpetual state of exchange with these underground reservoirs, sharing both naturally occurring substances and those introduced by human activities. This interchange can influence the groundwater's suitability for consumption and even the fertility of our soils for agricultural practices.

It is noteworthy that groundwater constitutes a staggering 99% of our accessible freshwater; a resource we have harnessed for our existence. Although concealed beneath the Earth's surface, its extraction and consumption have profound repercussions for both subterranean and terrestrial ecosystems. INRM serves as a tool enabling us to visualise groundwater's impact. Therefore, the use of groundwater is important in decisions aimed at improving water management.

Socio-Economic, Environmental and Cultural Inputs

For a socio-economic analysis, relevant information within the values of DSP and aligned with the UN SDGs could include:

1) **Water**
   - Access to Clean Water (SDG 6.1): Percentage with access to safe drinking water.
   - Water Quality Index (SDG 6.3): Measures water pollution levels.
   - Water Usage Efficiency (SDG 6.4): Evaluates water consumption.
   - Indigenous Water Management Practices (New Metric): Incorporates Indigenous wisdom on water stewardship.
   - Community Gathering Frequency (Cultural Indicator): Measures how often communities convene around water.
2) **Energy**
   - Access to Clean Energy (SDG 7.1): Percentage with access to reliable, clean energy.
   - Renewable Energy Adoption Rate (SDG 7.2): Share of renewables in the energy mix.
   - Energy Efficiency Index (SDG 7.3): Assesses energy consumption.
   - Indigenous Energy Solutions (New Metric): Integrates traditional energy methods.
   - Entrepreneur Ecosystem Strength (New Metric): Evaluates support for startups and small businesses.
   - Emissions Reduction Rate (SDG 13.2): Measures progress in reducing greenhouse gas (GHG) emissions.
3) **Biodiversity**
   - Biodiversity Index (SDG 15.1): Quantifies variety and health of species.
   - Ecosystem Health Score (SDG 15.4): Assesses the condition of ecosystems.
   - Conservation Efforts (SDG 15.5): Measures protected areas and projects.
   - Indigenous Conservation Practices (New Metric): Incorporates traditional land stewardship.
   - Sustainable Economic Practices (New Metric): Evaluates the economic activities' impact on biodiversity.
4) **Food**
   - Food Security Index (SDG 2.1): Measures food availability and access.
   - Nutritional Health Metrics (SDG 2.2): Tracks malnutrition and diet quality.
   - Sustainable Agriculture Practices (SDG 2.4): Evaluates eco friendly farming.

- Indigenous Food Systems (New Metric): Integrates traditional food cultivation and preparation.
- Economic Contribution of Agriculture (SDG 2.3): Measures agriculture's role in the local economy.

5) **Health**
   - Life Expectancy (SDG 3.2): Average years a person can expect to live.
   - Healthcare Access Index (SDG 3.8): Measures healthcare service availability.
   - Disease Prevalence Rates (SDG 3.3): Tracks major disease incidence.
   - Mental Health Well-being (SDG 3.4): Assessing psychological wellbeing.
   - Indigenous Healing Practices (New Metric): Includes traditional healing knowledge.

6) **Education**
   - Educational Attainment (SDG 4.1): Percentage completing various education levels.
   - Literacy Rate (SDG 4.6): Measures reading and writing proficiency.
   - Digital Literacy Score (SDG 4.4): Assesses digital skills.
   - Indigenous Knowledge Transmission (New Metric): Measures the preservation and transmission of Indigenous knowledge.
   - Storytelling and Cultural Learning (Cultural Indicator): Includes traditional story-sharing and community learning practices.

7) **Wealth, Defined as Prosperity Encompassing Economic, Social, Environmental and Cultural Well-Being**
   - Gini Coefficient (SDG 10.1): Quantifies income inequality.
   - Human Development Index (SDG 1.2, 8.6): Measures life expectancy, education and income.
   - Employment Rate (SDG 8.5): Percentage of working-age population employed.
   - Social Capital Index (New Metric): Measures community trust, social cohesion and engagement.
   - Resilience Score (New Metric): Evaluates the region's ability to withstand shocks and adapt to change.
   - Cultural and Community Indicators (Cultural Metric): Includes cultural preservation, community cohesion and collective identity.
   - Economic Drivers (New Metric): Evaluates the main economic drivers in the region.
   - Sustainable Economic Practices (New Metric): Assesses sustainable business practices and economic sustainability.
   - Emissions Reduction Rate (SDG 13.2): Measures progress in reducing GHG emissions.

These socio-economic, cultural and environmental indicators provide a well-rounded framework for sustainable development, aligning with the DSP approach and the UN SDGs and acknowledging the essence of each region's unique values and economy.

By collecting and analysing this socio-economic information, a clear baseline is established enabling informed decisions to address the region's unique challenges and leverage its strengths for a sustainable future. Engaging the whole team in the information gathering process from start to finish fosters a sense of ownership and shared responsibility. This approach ensures that everyone has a comprehensive understanding of the information and the insights it provides. Special attention should be given to identifying the most urgent needs, allowing the team to prioritise solutions that address immediate challenges.

In essence, a structured and inclusive approach, paired with a clear understanding of both global trends and regional needs, is crucial for achieving a sustainable future. It involves engaging the right people, immersing in local narratives and compiling a snapshot of the current situation to inspire future decisions.

- **Embedding the Right People**
  - Bringing in key interested parties ensures their perspectives are not just heard, but also embedded into the development. Involvement from the outset fosters an environment where key parties feel empowered, shaping the future with every decision made.
- **Delving into Community Narratives**
  - Collecting stories of the region holds profound significance for various reasons:
    o Cultural preservation for future generations
    o Sharing stories strengthens community ties and fosters a sense of belonging.
    o Stories shape the identity of the region giving residents a sense of pride in their roots, values and collective journey.
    o Insights into past challenges, successes and transformations offer lessons for the future.
    o Inspirational value. Tales of resilience, innovation and triumph can inspire others to overcome their own challenges and expand into possible opportunities.
- **Building a Socio-Economic Foundation**
  - Develop an initial DSP score, using visual tools like a spiderweb chart (see Figure 7.4) to illustrate the socio-economic landscape and pinpoint areas for expansion while highlighting the region's inherent strengths. An example is described in Chapter 5.

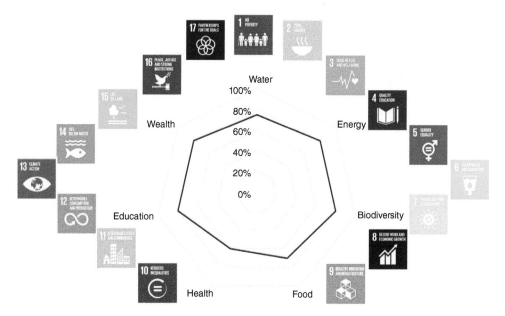

Figure 7.4 Socio-economic data for the region.

Market Study

The purpose is to anticipate and align with global trends and future market needs, addressing challenges such as climate change, pollution avoidance, energy transition and biodiversity conservation.

Innovative Opportunities for Sustainable Development

- **Green Energy Transition**
  - Exploring Solar Energy: From solar photovoltaic panels to concentrated solar power (CSP), harnessing the immense potential of an unlimited resource: the sun. Furthermore, the exploration extends to hybrid solutions where solar energy can be combined with geothermal, wind, tidal and hydropower sources wherever applicable.
  - Integration of Green Energy: It's not just about generation but integration; technologies such as CSP for direct desalination of seawater.
  - Exploring Carbon Sequestration and negative emission technologies. Drawing inspiration from natural processes like photosynthesis, such as artificial trees and mineral sequestration, to capture and repurpose carbon dioxide.

- **Water Management:** With only a small fraction of the Earth's water being fresh and accessible, innovations in water management are crucial. Strategies for water desalination, wastewater treatment and conservation:
  - Unlocking Freshwater resources, through direct desalination of seawater with CSP or hybrid options, play a vital role.
  - Reclaiming Wastewater for Reuse: From biological treatments to advanced filtration techniques.
  - Conserving Water: Be it large scale industrial processes or everyday tasks, strategies from recycling grey water to rainwater harvesting can make a tangible difference.
  - Improving Groundwater Management: Much of the world, particularly in arid regions, relies on groundwater for their needs. Improved mapping of the underground potential along with establishing a sustainable water balance can unlock arid regions.
- **Precision Agriculture:** With agriculture consuming 70% of global freshwater resources, the sector is ripe for innovation (Organization for Economic Co-operation and Development, 2023). Initiatives like conservation tillage for soil health, precision irrigation and scheduling for efficient water use and crop-livestock integration for energy conservation.
  - Conservation Tillage: By minimising soil disturbance to preserve soil quality and reduce energy consumption.
  - Smart Irrigation: Techniques like drip irrigation, paired with soil sensors, ensure water is delivered efficiently to plant roots, saving both water and energy.
  - Value Added Food Products: New ways to create value added products while retaining nutritional integrity such as high pressure processing, ultrasonic processing using high frequency sound waves and even 3D printing.

In essence, as we navigate the global market's complexities, the emphasis is on adopting sustainable strategies and innovative solutions that ensure prosperity while preserving our planet's resources.

Snapshot of the Needs of the Net Zero World

As the global market transitions towards sustainability, certain needs emerge:

- **Renewable Energy and Storage:** Harnessing and storing power from sustainable sources.
- **Sustainable Mobility:** Clean fuels, efficient public transport and green transport solutions.
- **Eco-Conscious Infrastructure:** Energy efficient materials, technologies and waste-minimising construction practices.
- **Circular Economy Solutions:** Innovative measures to ensure minimal waste and optimal resource use.

- **Climate Resilience:** Sustainable water management and especially groundwater recharge, flood prevention and preparedness for extreme weather conditions.
- **Carbon Dioxide and Methane Mitigation Technologies:** Emission elimination, carbon capture/storage solutions. Efficient carbon offset measures.
- **Sustainable Food Systems:** Environmentally friendly agriculture, waste reduction and alternative food sources.
- **Green Finance and Investment:** Funding channels exclusively for eco friendly projects and initiatives.
- **Smart and Sustainable Urbanisation:** Energy conserving cities, intelligent transport solutions and improved urban living standards.
- **Environmental Consultancy:** Expertise and guidance to adopt and implement sustainable practices.

Unmet Needs in Nature Positive Solutions

Nature offers a wealth of untapped opportunities for regenerative solutions:

- **Natural Carbon Sequestration:** Prioritising solutions like wetland regeneration afforestation and reforestation for effective carbon capture and long term storage.
- **Ecosystem Regeneration:** Initiatives to rejuvenate ecosystems, such as coral reefs, mangroves and peatlands, ensuring biodiversity protection and natural disaster mitigation.
- **Aquatic Restoration:** Elimination of pollution and sustainable management of aquatic ecosystems in both freshwater and marine environments.
- **Green Infrastructure:** Demand for nature inspired solutions, from green roofs to permeable pavements, ensuring urban resilience, reducing heat islands and enhancing overall liveability.
- **Regenerative Agriculture:** Embracing nature driven practices like agroforestry, permaculture and soil carbon sequestration ensuring sustainable farming without harming critical ecosystems.
- **Biodiversity Safeguarding:** Strategic efforts to conserve endangered species, biodiversity hotspots and critical habitats allowing ecosystems to thrive.
- **Natural Flood Management:** Harnessing the natural features of wetlands and river systems for sustainable land use practices for efficient flood risk management.
- **Climate-Aware Land Planning:** Shifting towards climate-resilient land planning and development that integrates nature inspired solutions to enhance resilience.
- **Blue Carbon Exploration:** Exploring and nurturing marine ecosystems like salt marshes and seagrass meadows, which sequester carbon and contribute to coastal protection.
- **Ecotourism Integrated with Regeneration:** Promoting experiences that not only showcase a region's natural splendour but actively integrate tourism into nature's rebirth.
- **Educational Outreach:** Amplifying educational outreach efforts through storytelling to immerse communities in a regenerative approach to nature, building awareness and support for sustainable practices.

In summary, these unmet needs underscore the immense potential for regenerative solutions rooted in nature. Addressing these needs not only tackles global environmental concerns but also yields significant socio-economic benefits for communities and regions, promoting a harmonious coexistence with our natural world.

The Journey of Seven Steps

To successfully navigate through these seven steps of the design journey, it is advisable to enlist a skilled facilitator to create an environment that encourages expansion and authentic communication, leading to transformation and fostering systems thinking.

Before the design it is essential to prepare each individual and the team, fostering a mindset of readiness and openness to change. This preparation lays the foundation for embracing new possibilities and driving transformation. Only once this preparation is complete can the team effectively develop the future for the region.

To change the thinking processes, the team should commence with the first two steps of preparation, fostering an environment that allows for bold envisaging, free from rigid constraints. This approach empowers the team to set their sights beyond their prior expectations and reference points. ***The key to successful transformation is for each individual to identify with their top strengths and those of the region***.

The seven steps guide a team's journey of discovering the potential to transform the region and themselves to shape the future of the region. Let's break it down (see Figure 7.5).

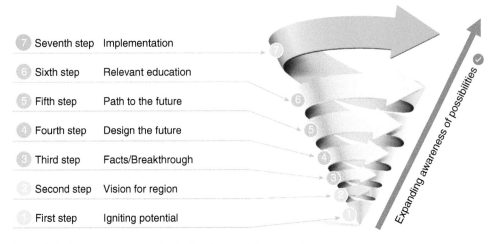

Figure 7.5 The seven steps of Designing Sustainable Prosperity.

Step 1 – Igniting Potential

There are numerous ways to ignite the team's potential, with practical exercises and activities being among the most effective methods. These immersive experiences should be structured to encourage expanding the awareness of possibilities. It is essential to let the team ask probing questions, experiment and embrace setbacks. When they see the limitations of their current thinking, they can start developing new ways of seeing and thinking. As they shed old ways of thinking, they become more adept at envisaging ambitious futures. Witnessing the positive outcomes of their new way of thinking ignites their motivation and determination. This shift in thinking lays the foundation for embracing systems level understanding and cultivating the conditions necessary for designing system solutions to create resilient and sustainable regions.

The method described below, guided by a skilled facilitator, serves as a good starting point for reflection after engaging in a cultural activity from the region.

- **Spark Curiosity:** To initiate the team's transformation journey and nurture the right mindset, engage in a regional cultural heritage activity. This experience is designed to cultivate curiosity, strengthen connections and encourage collaboration, exploration and discovery of new ideas. Select an activity that you believe will unite the team and then engage in these reflective questions:
  - What intrigued you most about the region's heritage?
  - How did this experience resonate with you personally?
  - What insights did you gain from it?
  - How did it impact your connection with the region and your team?
  - How can these insights drive transformation?
- **Envisage the Future:** Prompt participants to envisage a future that transcends existing boundaries. The objective is to nurture visionary thinking and push the envelope of possibilities.
  - If there were no limitations, what would you like to create for the region?
  - Describe your ideal future for the region.
  - How does leading without constraints feel for you and your team?
  - What actions or decisions do you make in this ideal future?
  - Can you capture this vision in a few words?
- **Explore Possibilities:** The example questions below will encourage the team to explore their core strengths, paving the way for challenging assumptions that may have limited their potential:
  - What strengths do you and your team possess?
  - What sets you apart?
  - What is the unique value you offer to the world?
  - How can these strengths complement each other to create a stronger team?
  - From the position of your strengths, boldly challenge old assumptions:
    - Can you identify any assumptions that have restricted your potential?
    - Are there any assumptions restricting your unique contributions?
  - How would leading from your strengths and embracing your unique gifts, look without old constraints?
- **Discover:** These questions will prompt the team to unleash the potential and embrace the narratives that will define their transformation to create the future for their region and lead to successful implementation.
  - What new possibilities do you see after challenging old assumptions?
  - How do you envisage the evolution of your leadership and its impact on both the team and the region?
  - How can you articulate the innovative narratives that will guide your transformation as a new leader?
- **Action (Navigating Success):** Design and execute small scale experiments aligned with your new ideas to set you on the path of impact. Example questions:
  - What is the smallest, most actionable step you can take right now to embrace the emerging possibilities?
  - How can you introduce these changes into your daily routine?
  - How can your team collectively support each other and celebrate progress?
- **Cultivate Skills:** These questions aim to promote ongoing skill development and resilience, enabling the team to maintain their progress even when facing obstacles and setbacks.

- What skills are essential to enhance your leadership for greater impact?
- How do you view setbacks as opportunities for enhancing your leadership impact and influence?
- What strategies can you implement to cultivate resilience within your team?

- **Zone of Expansion:** Continuously unfolding possibilities.

Empowered by step 1, the team will apply this newfound skill to unlock the region's potential, crafting a bold vision for the region.

Step 2 – Vision for the Region

Inspired by the region's stories and personal visions and the essence of the region, collaboratively craft a bold vision that leverages the region's strengths.

The following example questions will help bring this vision to life:

- **Storytelling:** Share the stories you have previously collected from the region's past and present, highlighting triumphs over challenges. Integrate Indigenous narratives that convey the essence of the land, sparking curiosity and exploration.
  - Which elements of the stories intrigued you the most?
  - How can these stories inspire your perception of the region's untapped potential?
- **Craft a Bold Vision:** Use the power of these stories and the region's essence to stimulate discussions that challenge conventional thinking. Envisage diverse scenarios for the region's future unburdened by current limitations.
  - How can these stories and the region's essence shape a bold vision?
  - Describe this vision vividly when fully realised.
  - What significant leaps can the region take without constraints?
- **Explore Possibilities:** Identify the region's unique strengths and unique purpose. Encourage an exploration of the disparities between this ambitious vision and the current state, all while leveraging the region's strengths.
  - Identify the region's standout strengths that can drive transformation.
  - Why does the region exist at its core and what is its intrinsic purpose?
  - What stands in the way of achieving the bold vision, considering the strengths of the region?
  - What needs to change to achieve the bold vision?
- **Discover:** Delve into uncharted territory and unleash possibilities.
  - What new opportunities emerge from the new way of thinking?
  - Paint a picture of this emerging reality aligned with the bold vision.
  - How does it feel to embrace these fresh possibilities?
- **Navigating Impact:**
  - What initial steps can be taken to set the path to success in motion?
  - How does it feel to take these initial steps towards realising the vision?

Having ignited the potential and forged a daring vision for the region, the team is empowered, prepared and ready to embark on the remaining steps of this journey.

Step 3 – Facts and Breakthroughs

Only when the team has dreamed about the possibilities for themselves and the region and are filled with curiosity and inspiration, they review the facts to validate their dreams. A breakthrough occurs when they realise that their dreams can become a reality.

- **Understanding Market Needs**
  – Identifying the market's most urgent needs, specifically those closest to a sustainable, net zero future that emphasises environmental conservation and promotes responsible actions. Key considerations include:
    o Identifying emerging green sectors and innovative clean technologies.
    o Anticipating potential technological advancements that could reshape market dynamics.
    o Addressing the most urgent societal needs.

  Seek to uncover the demands that will drive the future economy.

- **Exploring Regional Potential**
  Examine the previously prepared socio-economic baseline, environmental and cultural data and the INRM, allowing alignment of the identified market needs with the untapped potential of the region.
- **Establishing a Clear Baseline for the Region**
  – Define the region's boundaries.
  – Map the status.
    o Community
    o Population
    o Business base
    o Climate
    o Environment
    o Key players
  – Describe the region's infrastructure; linked by railways, major roads and international connections such as airports and seaports.
    o What is the region's digital connectivity status, such as fibre optics?
    o What foundational structures are in place, like hospitals, educational institutions and universities?
  – Develop a visual representation of the current system. It could also include a DSP score for the region (assessment of the region's performance with respect to water, energy, biodiversity, food, health, education and wealth). See example in Chapter 5 and in the socio-economic input section in this chapter.
- **Unveiling the Region's Potential by Examining the INRM**
  – Understand how the Earth was formed in the region by examining the INRM as shown in Figure 7.6. Gain insights into sub-surface, surface and climate potentials.
  – To extract valuable information from this visual representation of a region's potential, the following questions can be considered:
  – **Assess Sub-surface Potential**
    o What mineral systems are potentially present for sustainable extraction?
    o What potential does the sub-surface hold in terms of water, energy reserves, storage and other potential?
    o What is the region's water footprint?
    o Can the region leverage its geothermal profile for energy generation?
    o What indicators does geology provide about soil quality?
    o Is there any geological formation with significant storage potential?
    o Indicators for decisions on where to develop and where not to develop based on the knowledge of the sub-surface, such as protected areas for biodiversity or carbon sink and decisions for transmission lines, pipelines, rail, industry and housing.

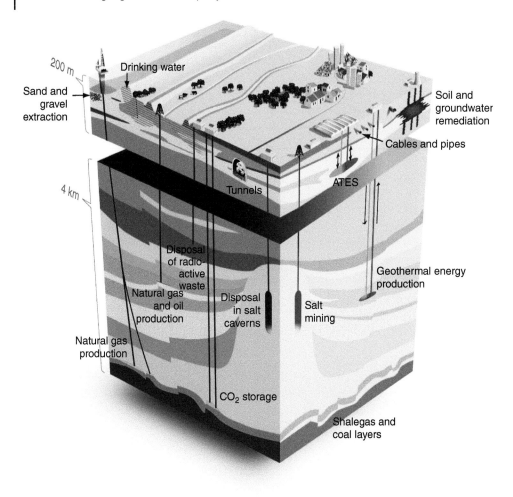

Figure 7.6 Layering of information to discover hidden potential in a region using an Integrated Natural Resource Model. *Source:* (Griffioen, et al., 2014 / with permission of Elsevier).

- **Assess Surface Potential**
  o Areas of high soil fertility to support agriculture.
  o Areas suitable for renewable energy installations like solar or wind farms.
  o Which regions are ecologically sensitive?
  o Which lands could be repurposed for economic development without compromising environmental integrity?
- **Climate Considerations**
  o How might the local climate conditions support or hinder the sustainable use of natural resources?
  o What climate-resilient practices could be adopted to safeguard and utilise these resources?
  o What potential climate risks does the region face?
  o Are there areas prone to natural hazards such as flooding, landslides, earthquakes or fires? How can development be planned to mitigate the risks associated with these hazards?
  o How can the region's development align with the UN SDGs?

- **Mapping Regional Potential**
  - Analysing the interactions between different potentials, identifying synergies and constraints
  - Developing a visual representation of all potentials, laying out how each key element can interconnect within the system.
  - Align the region's strengths and resources with the market needs. Start by highlighting the gaps and opportunities in the market:
    o Where does the region fit in?
    o What can it offer?
  - Relate demographic strengths to market needs:
    o What are the unique skills the population possesses?
    o What demographic patterns characterise the region?
    o How can the region's demographic profile be leveraged?
    o Integrate views on harnessing the power of the region's diverse talents and culture to meet market demands.
  - Highlight the potential of the region (natural resources and overall regional strengths).
    o What stands out as the region's primary strengths?
    o Elaborate these ideas in sync with market needs and demographic profiles.
  - Relate the region's infrastructure (transportation, utilities, technological backbone, educational facilities, etc.) to market needs.
  - Evaluate proposed value ideas:
    o Determine, how the region's strengths can meet the unmet needs of the identified markets.
- **Readiness for the Next Phase**
  - The team has now achieved a breakthrough in expanding their awareness of possibilities for themselves and the region.
  - Market needs are aligned with the untapped potential of the region.
  - Critical regional needs, core strengths and strategies to address challenges are conceptualised.
  - The team is energised and ready to co-create the future in the next step.

Step 4 – Solutions System

This step presents the challenge of crafting a blueprint for a prosperous future. The previous phases primed the team, liberated their minds from constraints and prepared them for this critical step.

- **Designing a Sustainable Solutions System**
  - Address future demands for environmental protection and climate change mitigation, emphasising biodiversity and ecosystem health.
  - Focus on green energy solutions, addressing water challenges and enabling agriculture and innovation.
  - Build high level business cases, prioritising options using the UN SDG indicators.
  - Key questions:
    o Envisage sustainable enterprises aligned with regional potential.
    o Devise strategies to attract these enterprises.
    o Create enticing incentives for potential investors.
  - The process of discerning unmet needs, unlocking regional potential and conceptualising sustainable enterprises is dynamic and iterative.

- **Systems Design Phase Structure**
  - Develop a Shared Vision: Integrate environmental, social and economic sustainability into the vision of the new system.
  - Define New System Components: Identify the key components and elements needed to bring the shared vision to life (e.g., energy, water, agriculture, mining and infrastructure).
  - Design Integrated and Synergistic Solutions: Create efficient and waste reducing solutions within the new system.
  - Create Scalable Frameworks: Develop adaptable structures for the new system, capable of accommodating future changes.
  - Optimise Resource Use: Promote renewable energy, water conservation and sustainable practices within the new system.
  - Adopt Technologies: Adopt technologies that enhance system efficiency and resilience in the new system.
  - Provide Education and Training: Offer training programs to build local capacity for managing and maintaining the new system.
  - Aligning Policies: Develop policies that support sustainability and innovation within the system.
- **Facts Reassessment:** We embark on a journey to realign our understanding of the region's potential and redefine our unique value propositions. Our process unfolds as follows:
  - Reassess the facts to create alignment and define unique value propositions.
    - Delve deep into the knowledge of the region to unearth its hidden potential.
    - Identify and highlight the region's most compelling strengths.
    - Address the region's primary challenges by capitalising on its inherent strengths.
    - Identify the specific markets that align with the region's unique value proposition.
  - Identify additional challenges or opportunities.
    - Identify novel solutions that arise in response to these new challenges and opportunities.
  - Tap into untapped markets:
    - Identify untapped markets and unmet needs that can benefit from the region's offerings.
    - What other high value the region can offer for the unmet needs of the markets?
  - Propose high value strategies:
    - Develop strategies to deliver high value solutions to these identified markets.
- **Water, Energy and Infrastructure Analysis Example**
  In the pursuit of regional sustainability, it is vital to design solutions that address the unique challenges and opportunities within a given region. While this example may not fit every region, it serves as a practical illustration of the system design process:
  - **Water Management**
    - Evaluate the entire regional water cycle, focusing on efficient and responsible water resource management.
    - Tackle issues related to both water scarcity and abundance.
    - Ensure the availability of abundant, affordable and sustainable energy to support water solutions.
  - **Energy Transition**
    - Examine the region's current energy matrix.
    - Facilitate a transition towards green, abundant and affordable energy sources.
    - Acknowledge that clean energy is a driving force behind regional development for sustainability.

- **Exploring Additional Opportunities**
  - o Beyond water and energy, explore other possibilities, often beginning with agriculture.
  - o Embrace precision agriculture, emphasising resource efficiency and biodiversity.
  - o Recognise agriculture's role in supporting food security and alignment with other sectors such as value added food products and infrastructure.
- **Infrastructure and Waste Management**
  - o Assess critical infrastructure such as transportation, telecommunications and logistics as they are essential for efficient industries.
  - o Evaluate waste management solutions to ensure sustainability.
  - o Recognise that an integrated approach to infrastructure and waste management aligns with sustainable goals.
- **Building a Knowledge-Based Economy**
  - o Develop intellectual property and expertise within the region.
  - o Strive to become a centre of excellence in specific technologies or processes.
  - o Promote innovation as a catalyst for sustainable development.
- **Exploring Synergies and Addressing Environmental Impact**
  - o Assess potential synergies between mining, agriculture, water and energy resources.
  - o Consider the environmental impact of these sectors on local communities and ecosystems.
  - o Develop integrated solutions to enhance sustainability and resilience.
- **Aligning with UN SDGs and Prioritisation**
  - o Create high level business cases for the top solutions.
  - o Map the impact of these solutions on relevant SDGs, with a focus on primary and secondary SDGs.
  - o Score ideas based on their alignment with SDGs and potential impact.
  - o Rank these ideas according to their scores to prioritise implementation.
- **Delineating Value**
  - Define the value propositions of the region's services, technologies and products and offerings.
  - Highlight the competitive advantages that set the offerings apart.
  - Identify unique differentiators such as innovative elements, intellectual property or unique approaches that are not readily available elsewhere.
- **Investor Types and Strategies**
  - Reflect on the following questions:
    - o Who are our ideal investors and why are they the perfect match for our offerings?
    - o What differentiated solutions address the specific challenges of these investors?
  - Envisage ideal investors for the region and their potential interests.
  - Identify intellectual property opportunities to bolster the knowledge-based economy.
  - Explore potential alliances to further refine and realise the top ideas.
  - Create investment propositions showcasing unique regional opportunities.
- **Mapping the Network of Ideas**
  - Connect ideas to outline solutions aligned with the regional strengths and market needs.
- **System Solution Synthesis**
  - Refine the visual representation of the prioritised solutions.
  - Create multiple models projecting system evolution under various conditions.
  - Determine key intervention points with significant positive ripple effects.
  - Integrate findings and strategies into a cohesive system solution.

  – Ensure flexibility and adaptability for future expansion, innovation and sustainable resource use.
  – Prepare for implementation, monitoring and continual reassessment.
- **Design Outcomes**
  – Identified system solutions, including long-lasting enterprises (services, products and technologies) that are ideally suited for the region.
  – Prioritised rapid and decisive actions to secure early victories and propel the team towards success.
  – Identified necessary changes for creating an environment conducive to future opportunities.
  – Developed attractive investment packages.
  – Set a trajectory for successful investment within five years post planning.

This approach aims to create a roadmap for the region's sustainable development, ensuring alignment with conservation efforts and strategies for climate change mitigation.

Step 5 – Path of Impact

Upon selection and prioritisation of solutions, a roadmap is created for the business cases for maximum impact.

This roadmap follows a phased approach, with clear steps and milestones, to ensure achievable progress, as shown in Figure 7.7.

- **Evaluating Impact**
  – Assess the ripple effect of each solution.
  – Identify propositions with transformative and sustained impact.
- **Prioritisation**
  – Segment solutions into categories (Figure 7.7):
    o Short Term Actions: These are initiatives that can be easily implemented, yielding immediate benefits and laying the path for success.
    o Medium Term Strategies: Building on short term successes, these solutions require a more intricate setup and a longer gestation period but promise a substantial impact.
    o Long Term Vision: These are your landmark projects, which, while taking time, shape the horizon of your envisaged future. They represent sustainable and transformative change.
  – Prioritise solutions based on impact and timeframes.

Figure 7.7   Phased approach roadmap for the future. *Source:* pingebat/Adobe Stock Photos.

- **Integrated Roadmap Development**
  - Weave solutions into a cohesive roadmap.
  - Define actions, milestones, checkpoints and feedback loops.
  - Ensure alignment with the overarching vision.
- **Key Parties' Collaboration**
  - Engage key parties throughout the roadmap development.
  - Secure buy-in, feedback and collaborative efforts to ensure effective implementation.
- **Flexibility and Adaptability**
  - Build provisions for periodic reviews.
  - Allow course corrections, adaptations and integration of emergent solutions and technologies.
- **Monitoring and Evaluation**
  - Incorporate performance metrics and indicators for ongoing monitoring and refinement.

Step 6 – Relevant Education

In anticipation of a sustainable future, the team envisages an adaptable, dynamic education system that nurtures creativity and curiosity from childhood to higher education. This step involves reviewing the necessary changes in the education system to equip individuals with the skills and tools to support future diverse enterprises.

Key Elements

- Understand the current education system and cultural and ancestral aspects.
- Define the skills needed for long-lasting enterprises to meet future needs.
- Plan to augment further education to support new skills and capabilities required for the future.
- Plan for retraining the current workforce when necessary to align with the future. Completion of the design phase enables anticipation of future skill requirements, facilitating the retraining plan.
- Develop and enhance leadership skills and capabilities.
- Adapt children's education to unleash their high potential. The education chapter provides an overview and deeper insights.
  - Foster an environment that encourages creativity as a way of life. Use Beau Lotto's 'i,scientist' (Canadian Institute of Mining, Metallurgy and Petroleum, 2022) method to adapt children's education and unleash their potential.
- Lay the foundation for a knowledge-based economy.

Step 7 – Action and Maintaining the Momentum

This step prepares the environment to implement the DSP plan, transitioning into a dynamic resource that requires periodic reviews. It encompasses short, medium and long term actions and ensures changes are made when appropriate. The objective is to create an environment for action, maintain momentum and continually embrace innovation.

- **Phases of Implementation**
  - Break down the implementation into manageable phases – short, medium and long term.
  - Ensure real-time adjustments and continual progress towards set milestones.
- **Creating Innovation Hubs**
  - Establish centres of excellence where 'out of the box' thinking is not just encouraged but celebrated.

- Encourage multidisciplinary teams to challenge the status quo and pioneer differentiated offerings.
- Diversify innovation hubs with experts from varied domains.
- Cross-pollinate knowledge for holistic and groundbreaking solutions.
- Regularly update skillsets to remain equipped for innovation.
- Deliver solutions that meet urgent market needs.
- **Living on the Edge of Innovation**
  - Aim to set global trends rather than just follow them.
  - Position the region at the forefront of emerging creativity.
- **Key Parties' Integration**
  - Embed more key parties as co-creators, not just contributors.
  - Integrate their insights, experiences and foresight into the fabric of innovation.
  - Invite key parties also as co-creators.
  - Continually gather feedback from key parties.
  - Ensure alignment with market needs and adapt to changing dynamics.
- **Sustainability as a Culture**
  - Instil sustainability as a core cultural value.
  - Align everyone with the mission of creating differentiated, sustainable offerings.
- **Championing the Cause**
  - Identify and nurture more champions within the enterprises who can drive the mission with passion.
  - Ensure momentum and motivation remain high throughout the implementation journey.
- **Regularly Measure and Celebrate Milestones**
  - Implement a tracking mechanism to measure progress against milestones.
  - Celebrate achievements to boost morale and reinforce commitment to the mission.
  - Recognise and acknowledge the efforts and successes of individuals and teams.
  - Use celebrations as opportunities to maintain enthusiasm throughout the journey.

In essence, the path to sustainable innovation is a cyclical process of creation, implementation, evaluation and regeneration. With commitment and an agile approach, success with sustainable, differentiated offerings becomes a reality.

Implementation

This book primarily focuses on the planning and preparation for implementation, recognising that each region's implementation strategy will be unique. It does not delve into the specific execution details, as those will be tailored to each region's specific needs and potentials.

The plan outlined in this method chapter serves as a dynamic resource that can be updated after reaching each milestone. This holistic approach to preparing the environment sets the stage for efficient execution, ensuring that the region is well equipped to transform its vision into reality with a focus on action.

Implementation involves a series of key steps:

1) **Engagement with Potential Investors**
   - Collaborate with potential investors who are aligned with the region's vision and values.
   - Establish partnerships and secure financial support to bring sustainable solutions to life.
   - Advance the business plans developed in the planning process.

2) **Preparing the Environment for Efficient Execution**
   - Implement changes in the education system, training programs and educational infrastructure as needed to nurture future-ready talent.
   - Collaborate with policymakers to develop and align regulations and policies that support sustainable development.
   - Invest in the necessary infrastructure required to support the new enterprises.
   - Establish robust systems for monitoring progress and evaluating outcomes.
   - Regularly assess the impact of initiatives, gather feedback and adjust if needed to stay on course.

3) **Sustainable Development: Maintaining Momentum and Seeking Further Innovation**
   - Embed sustainability into the core values and practices to ensure the continual development of the region.
   - Seek opportunities for further innovation and positive impact on the region.
   - Stay at the forefront of creativity to continually push boundaries and set new standards in sustainable development.

Encouraging the establishment of vibrant, long-lasting enterprises that enhance the environment, quality of life, health and well-being should be the goal for resource rich regions.

To truly grasp the nuances of the DSP method, immerse yourself in the Chapter 8 which presents high level case studies. These narratives elucidate the DSP journey, highlighting a diverse array of regional challenges and their resilient sustainable solutions.

References

Canadian Institute of Mining, Metallurgy and Petroleum. (2022, October 14). *I, scientist program: giving young people the skills needed to thrive in an uncertain world*. Retrieved from Canadian Institute of Mining, Metallurgy and Petroleum: https://www.cim.org/news/2022/i-scientist-program-giving-young-people-the-skills-needed-to-thrive-in-an-uncertain-world/

Griffioen, J., van Wensemd, J., Oomesd, J. L., Barendse, F., Breunesec, J., Bruining, H., Olsthoorn, T., Stams, A. J., & van der Stoeli, A. E. (2014). A technical investigation on tools and concepts for sustainable management of the subsurface in The Netherlands. *Science of The Total Environment*, 485:810–819. Retrieved from https://energy.danube-region.eu/wp-content/uploads/sites/6/sites/6/2021/03/Energy_storage2021FINAL_03.25.pdf

Hiam-Galvez, D., Prescott, F., & Hiam, J. (2020). Designing Sustainable Prosperity "DSP": A collaborative effort to build resilience in mining producing regions. *CIM Journal*, 11 (1):69–79.

Organization for Economic Co-operation and Development. (2023). *Managing water sustainably is key to the future of food and agriculture*. Retrieved from Organization for Economic Co-operation and Development: https://www.oecd.org/agriculture/topics/water-and-agriculture/

8

Case Studies
Doris Hiam-Galvez[1], Lily Lai Chi So[2], and Richard Blewett[3]

[1] Hatch Ltd, Designing Sustainable Prosperity, Vancouver, British Columbia, Canada
[2] Hatch Ltd, Innovation and Investments, Mississauga, Ontario, Canada
[3] GeoSystems Consulting, Geoscience, Royalla, New South Wales, Australia

Introduction

The natural resource industry is cyclical, characterised by commodity price volatility and economic growth tied to the life cycle of the mine. This often leads to boom and bust cycles. By applying the principles of DSP, a region's economy can be designed and developed with resilience in mind, moving away from the traditional cyclical nature of the natural resource industry.

Before we step into case studies that demonstrate the implementation of the DSP method, let us first look at the diversification of Colorado (Figure 8.1). This development of the Denver area predates DSP but illustrates the successful application and implementation of some of the elements of the DSP method. Denver is a success story that began with an inspired champion who started a collaborative movement, drawing in many more champions to steer the economy away from 'boom and bust cycles'.

Denver's economy historically followed the 'boom and bust' cycles; for example, make money during the boom, ride the 'bust' with the money made in the 'boom' and wait in hopeful anticipation for the next 'boom'. In the late nineteenth century, Denver was built on railways, minerals and ranching. After the gold rush ended in the early 1860s (Encyclopedia Staff, 2023), a growing manufacturing industry revived the town. In the twentieth century, preceding World War II, Denver, situated far from either coast, became a hub to support the war efforts, attracting the growth of military and federal activity in the aerospace industry. After the war, Denver and its adjacent region saw a boom in the oil and gas industry as the region was close to oil fields. Huge federal and private spending resulted in rapid expansion of the city. But when the energy crisis hit in the 1980s, the overbuilt region saw a mass exodus and became stagnant. This was a wake-up call for Denver.

With the continual rise and fall of the historical mining and oil and gas industries, Denver deemed it necessary in the 1980s to diversify its economy to reduce its dependence on the cyclical nature of these industries, that is, not to wait around and count on the next boom to revive the city. In applying the concept of DSP in the analysis of the development of Denver, we identified several critical elements that enabled the change:

- Key leaders with the right vision and skills to motivate and drive the efforts.
- Cooperative relationships among the business sector, local cities and economic development staff.
- Commitment to collaboratively build the right environment.
- Infrastructure – the key was a larger international airport with scope for expansion (New Denver International Airport).
- They already had core businesses to build on (aerospace and telecommunications).
- Accessibility.
- Education – to support the diverse economy, all levels and systems of education including basic skills relevant to the businesses were available.

Figure 8.1 Location of Colorado.

To build a resilient economy, one needs to look at key elements that enable diversification beyond extraction of natural resources. The Metro Denver Economic Development Corporation raised funds to commission market studies to develop a plan to diversify their economy. It was a regional plan where all institutions, public and private, agreed to work together to achieve the greatest good for the greatest number of individuals (Metro Denver Economic Development Corporation, 2022).

Rules of behaviour were fostered where everyone wanted to be part of the group for the regional good. Denver and its surrounding counties agreed to not compete against each other for new business and talent. Agreement was struck on regional tax rates and recruiting key industries like technology companies. Areas dedicated to specific industries, such as aerospace, were built. Aviation, renewable energy, bioscience and telecommunications were specifically targeted.

The key to success was modernisation of infrastructure. The new Denver International Airport was built together with all the supporting infrastructure. Colorado had a good base on which to build. Many companies were already present and served as the foundation for growth. First-class universities were already established. The location was convenient and considered a desirable place to live, making it easy to attract the necessary talents and people to support a diversified economy. This does not detract from the fact that a good effort was made with all the right features in the program.

The journey that Colorado went through in the past three decades demonstrated the successful building of a diversified economy. Figure 8.2 shows the change in employment for Colorado between 1991 and 2018 and clearly demonstrates the robustness enabled by a diversified economy – drastic employment changes in volatile and cyclic 'boom and bust' industries such as extractive resources are balanced by high-tech industries which are less susceptible to market volatility. Today, Denver's economy includes strong, growth oriented and high-tech industries while maintaining its reputation as a hub for the more cyclical extractive resources industries, coupled with a convenient and accessible location adjacent to the Rockies. Calgary has a similar situation to Denver's in that it is currently mainly dependent on a few cyclical industries and could apply the DSP principles to create a more resilient region. Economies are dynamic and many other regions need an intended effort to increase resilience of their regions. DSP principles when applied need regular reviews, so the process is continually evolving. Denver has recognised this and is ready for a review with a view to maintaining economic resilience.

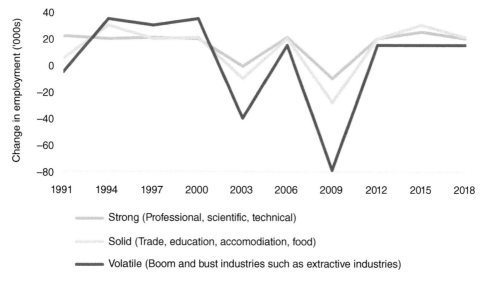

Figure 8.2 Change in employment for Colorado between 1991 and 2018. *Source:* Adapted from Metro Denver Economic Development Corporation (2018).

The case studies in this book provide a high level overview, demonstrating how the DSP method can be adapted to meet diverse regional challenges. These preliminary assessments offer a glimpse into the potential of DSP in various settings. As you delve into these case studies, it will become evident that DSP is a flexible approach, capable of yielding tailored solutions. The method chapter goes deeper into the planning process, outlining how DSP can be systematically applied. While this book primarily focuses on the planning aspect, it is important to remember that the implementation phase will be distinct for each region, reflecting their specific needs and potential.

All these case studies align with the United Nations Sustainable Development Goals (UN SDGs) as the method is crafted to encompass all facets of sustainable development to create resilient regions in natural resource rich regions.

The following studies are described:

1) **Peru South:** Solar-powered economy
2) **Caring for Country:** From dust to wonder: Uncovering hidden potential
3) **Northern Ontario's 'Ring of Fire', Canada:** Nature positive solution
4) **Northwest British Columbia's Canada**, Low carbon solution hub
5) **Quebec Region, Canada:** An integrated green energy system
6) **Peru's Sechura Desert:** A circular industrial cluster

Solar Powered Southern Peru Economy

Alignment to UN SDGs 1, 2, 3, 4, 6, 7, 8, 11, 12, 13, 14, 15, 17

Summary

In Southern Peru, a preliminary DSP study reveals the potential for innovative processes and technologies that can harness regional resources, leverages the region's mining and metallurgical knowledge and utilises existing labour and trade skills to provide sustainable and environmentally friendly access to scarce water resources.

This opportunity, as depicted in Figure 8.3, leverages a combination of solar and geothermal energy resources, along with innovative technologies, to address water scarcity and drive economic diversification. This region, which is rich in copper and vital for the global energy transition, faces water scarcity. Despite being a water scarce region, its proximity to the ocean, excellent workforce adept at implementing new technologies and fertile soil position it as a strategic location for economic diversification.

By integrating advanced concentrated solar power (CSP) with geothermal energy for seawater desalination and harnessing sequential crystallisation processes for mineral recovery from the concentrated brines. By implementing advanced direct seawater desalination with solar power, the region could address its water scarcity challenges, thus enabling the development of precision agriculture and value added food products. This will allow Southern Peru to become a centre of excellence in making these complex technologies economically viable and then export this know-how globally.

Figure 8.3 Peru's southern region and indicative solar irradiance, where red is high and green is low.

Current State

Peru's economy heavily relies on mining, particularly copper (Figure 8.4). However, impending water shortages and climate change pose significant challenges. The region's robust solar irradiance and geothermal potential offer unique opportunities for sustainable energy and water solutions opening the door for other sustainable enterprises to be developed.

Peru is the second-largest copper producer worldwide (Statista, 2022a), with the mining industry accounting for over 30% of employment (Oxford Business Group, 2022; Trading Economics, 2022) and agriculture for nearly 30% (Statista, 2022b). As decarbonisation and food shortage concerns rise, the demand for metal and agricultural products grows. Peru faces two significant challenges: the local economy's survival, post-mining resource depletion and an impending water shortage within a decade. Agriculture accounts for nearly 90% of the water demand (Miller & Sweigart, 2019) and with increasing industrial activity and population needs, water demand is expected to surge. Yet, Peru is one of the most water stressed countries globally, with water availability projected to decline due to climate change's impact on Andean snow cover (Hiam-Galvez, Gleming, & Osores, 2012).

$64.7B

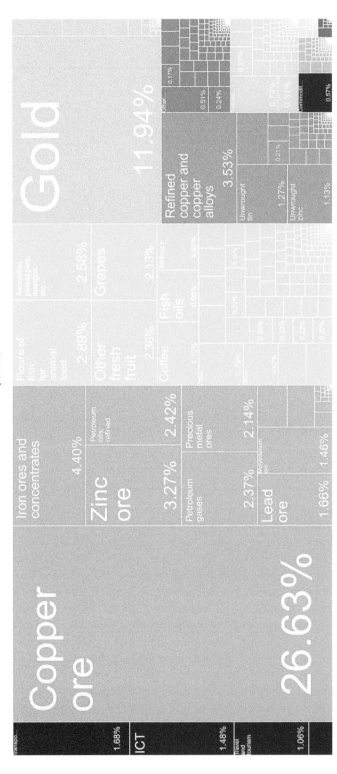

Figure 8.4 Peru's export map in 2021. *Source:* Atlas of Economic Complexity (n.d.) / https://atlas.cid.harvard.edu/explore.

Despite being identified as one of the world's most vulnerable countries to climate change impacts by the Tyndall Centre for Climate Change Research, the region offers immense renewable energy potential (Bebbington & Williams, 2008; Hiam-Galvez, Gleming, & Osores, 2012). Peru has the highest global solar irradiance levels (Figure 8.5) and its Pacific Ring of Fire location gives it a geothermal potential equivalent to 50% of its current electricity production (Richter, 2020).

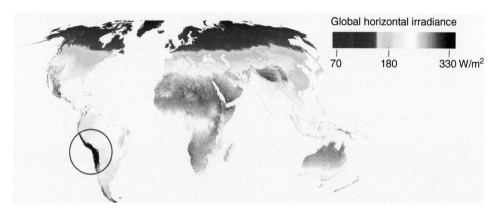

Figure 8.5 Global mean solar irradiance. *Source:* Vaisala (2017) /with permission of Vaisala.

Untapped Potential

Peru's natural endowment presents a tapestry of opportunities waiting to be woven into the fabric of its future. These include:

1) A diverse ecosystem with rich biodiversity suitable for eco-tourism and the development of natural products.
2) With the world's best solar irradiation potential (Figure 8.5), making it ideal not only for desalination but also for electricity generation and agricultural applications.
3) Geothermal energy potential, due to its location on the Pacific Ring of Fire, opens doors to electricity generation, heating applications and even wellness tourism.
4) Proximity to the sea offers opportunities for seawater desalination, fisheries and maritime transportation.
5) The fertile soil requiring only water to produce premium produce such as olives, grapes and more. These products can be transformed into high value added food products.
6) A rich cultural heritage that offers opportunities for cultural tourism, traditional arts and crafts and the promotion of traditional agricultural practices.
7) A highly skilled, educated and resourceful workforce with the ability to implement and improve new technologies. This human capital is a critical asset for the development of innovative solutions to the challenges faced by the region.
8) A mix of developed and semi-developed regions offers opportunities for synergies and the sharing of best practices and resources.
9) Universities in the region that offer opportunities for research and development and the training of skilled personnel.
10) An opportunity to play a critical role in the global energy transition by supplying copper and other critical metals and minerals. This positions the region as a key player in tackling climate change.

Challenges and Risks

Nevertheless, a set of challenges and risks loom on the horizon:

1) Water scarcity and quality is a major challenge in the region, particularly in arid areas.
2) Geographic barriers complicating access and infrastructure development in mountain areas.
3) Vulnerability to impacts of climate change in precipitation patterns, temperature increases and extreme weather events.
4) Widespread communities in remote, mountainous areas present challenges for the provision of services and infrastructure.
5) High levels of poverty.
6) The region has a rich biodiverse ecosystem that needs to be preserved. Economic development activities need to be carefully planned to ensure the sustainability of the ecosystem.

Potential Future Offered by DSP

Southern Peru could become a global leader in sustainable energy and water management. The key to this lies in the integration of advanced CSP and geothermal technologies unlocking efficient and economical seawater desalination capabilities. Additionally, the region can harness technologies to recover valuable minerals from concentrated brines produced from desalination. This transformative approach promises to address Southern Peru's water scarcity challenges while simultaneously opening the door to economic diversification, as access to water and energy lays the foundation for other opportunities.

Advanced direct seawater desalination powered by solar energy could pave the way for smart agriculture and high value added food products. This positions Peru as a shining example of how to make these complex technologies economically viable, not just for the region but also on a global scale. Beyond mitigating water scarcity, it offers a gateway for other innovative processes, ushering in a new era for the region. In essence, Southern Peru is uniquely poised to evolve into a global hub for complex energy solutions and the advancement of sustainable practices.

1) **Innovative Technologies**

Leading the charge in this transformation is the pioneering Solar Dome Thermal Desalination technology. This innovative technology being piloted at Neom (Solar Water Plc, 2023) and developed by the UK's Solar Power and Cranfield University (Sansom, Patchigolla, Jonnalagadda, & King, 2020), harnesses solar energy to produce one million litres of clean water per day while achieving carbon neutrality. A sketch of the Solar Dome technology is shown in Figure 8.6.

This eco friendly approach significantly deviates from conventional reverse osmosis (RO) desalination technologies that often entail high energy consumption and environmental drawbacks. RO uses large quantities of energy to pump the water through membranes and a high percentage of salt and dissolved valuable materials is rejected (Poirier, Al Mhanna, & Patchigolla, 2022). The solar dome technology uses direct evaporation and distillation with solar power.

Central to this breakthrough is a multi-effect distillation process that efficiently reuses latent heat and minimises the need for cooling water.

The concentrated brine can be processed by an innovative multi-crystallisation process (Poirier, Al Mhanna, & Patchigolla, 2022) to recover high purity salts and a sophisticated heat recovery system, ultimately striving for a zero liquid discharge goal.

Figure 8.6 Concentrating Solar Power technology for desalination of seawater. *Source:* Love Employee/Adobe Stock Photos.

This thermal desalination technology can also be implemented as a hybrid, combining solar and geothermal energy to create a cost effective, energy efficient and environmentally friendly freshwater production facility. This hybrid approach ensures stable power generation and minimises the need for thermal energy storage. This ensures uninterrupted desalination and energy availability.

These thermal systems are also being tested in small and medium scale CSP plants, further validating their effectiveness (ENEA, 2017). Additionally, the system can integrate a heat recovery mechanism for an absorption chiller and desalination from CSP steam power plants, as exemplified by its pilot implementation in Egypt (Bruch, et al., 2020).

In summary, this integrated system, encompassing CSP, thermal energy storage, geothermal energy, thermal desalination and brine processing technologies, offers a comprehensive solution to freshwater scarcity and lays the foundation for other complex processes. It represents significant progress towards carbon neutrality, cost effectiveness, zero liquid discharge and uninterrupted production.

2) **Leveraging Regional Workforce Expertise**

To catalyse the economic viability of advanced solar water technologies and expand the expertise into other industrial processes, the region can harness its workforce's specialised knowledge. With a deep rooted understanding of high temperature processing and equipment operation from mining and copper smelter industry, these skilled workers possess a unique advantage that will anchor the success of this new technology. Their seamless transition into roles within the solar and geothermal thermal processes and water treatment can be instrumental in driving innovation and progress with these complex industrial processes.

This transition extends beyond technology adoption. It signifies a paradigm shift, where the region becomes recognised for its remarkable capacity to make complex systems work. This reputation for problem solving and innovation becomes the foundation upon which the region diversifies, with solar water technologies at its core.

Furthermore, the education system needs to be adapted to support the development of the skills and tools needed to support these complex industrial processes; all levels of education can be developed around solar and geothermal technologies and desalination of seawater, brine crystallisation and value added food products.

System Solutions

The cyclic nature of the mining and metals industry can be absorbed by the more robust energy and water economies that DSP envisages. With abundant clean water and affordable clean energy, the agricultural and food sector can flourish, encompassing not only fresh produce but also high value added food products. This supports a sustainable future.

To embark on this journey, Peru can initiate partnerships with leaders in the field and continue developing the technology to make it more viable, thereby becoming developers and not just purchasers of the technology. There is therefore an opportunity to focus on improving the hybrid solar and geothermal technology to use as an economic source of energy for direct seawater desalination.

The visual representation in Figure 8.7 aptly illustrates how CSP desalination serves as the nucleus for a diversified economy. Water and energy serve as catalytic elements to produce agricultural goods for the future economy and are represented in the lower section. The copper industry and its associated skills stand as essential facilitators for the region's resilience. The expertise in high temperature processing, positioned at the top, fuels the development of a pivotal new skill set: the capacity to render things viable and operational.

Figure 8.7 envisages a future where Peru perfects the technology of direct seawater desalination using solar power, establishing itself as a global leader and centre of excellence for this know-how.

Finally, Figure 8.8 outlines the next steps and a potential trajectory designed for maximal impact for Southern Peru.

Conclusion

Peru's Southern region has the potential to lead a sustainable economic transformation, leveraging its natural resources and their ability to adapt and evolve complex industrial processes. Through the DSP approach, characterised by integrating solar and geothermal power for direct seawater desalination, Peru can set a global example of resilience and prosperity in harmony with the environment. Clean energy and water will lay the foundation for other sustainable enterprises such as other industrial processes, smart agriculture and beyond, creating opportunities for the development of value added food products for the global market. As a centre of excellence in direct seawater desalination powered by solar and geothermal energy, the region can then export this invaluable knowledge to the world.

We have only taken the first steps in the DSP approach by discovering what possibilities lie ahead and what the design basis of the future ecosystem could look like. We would need to engage with the key parties in the region to co-create this future.

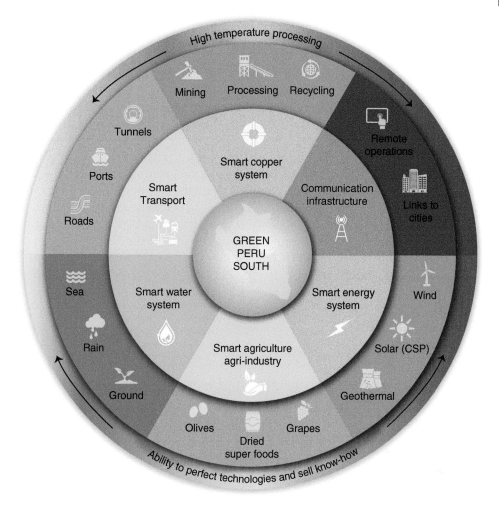

Figure 8.7   Green Peru South.

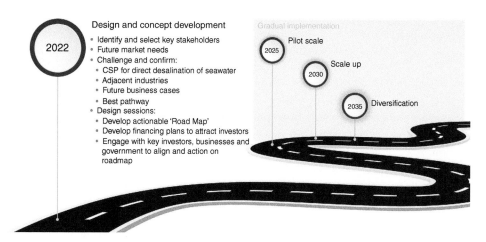

Figure 8.8   Possible path towards Peru's Solar-Powered Economy of the future.

Caring for Country – From Dust to Wonder: Uncovering Hidden Potential

Alignment to UN SDGs 1, 2, 3, 4, 6, 7, 8, 11, 12, 13, 14, 15, 17

For the past 60,000 years, Aboriginal peoples have cared for their ancient lands on the edge of the Great Victoria Desert, Australia (Figure 8.9). This land was not always desert. When the climate was wetter, the region's lakes and rivers ran with freshwater. Megafauna, like giant kangaroos and wombats, roamed the land and evolved with the unique marsupials of Australia. There were periods when the climate was even drier than today, such as during the great ice ages that covered much of the Northern Hemisphere. It was very arid and windy and great salt lakes developed and sand dunes rolled across the flat landscape. Surface water was scarce and the plants, animals and people contracted to limited refugia. These climate forces and landscapes have created the unique ecosystems and biodiversity we see today.

Figure 8.9   A typical 'big sky' landscape where the horizon seems to go forever.

Four main indigenous group's ancestral lands meet around the present day towns of Leonora and Laverton in the Eastern Goldfields region of Western Australia (Shire of Laverton, 2022; Shire of Leonora, 2022). To the east stretches the Great Victoria Desert – a vast region of wilderness and beauty. To the south lies the large regional centre of Kalgoorlie – famous for its gold and nickel mines and the gateway to the wilds of the Southern Ocean. A day's drive to the west lies the cosmopolitan capital city of Western Australia, Perth.

The climate is arid (semi-desert to desert) making it hot in summer and cool in winter. Natural hazards are few in comparison to other areas: a stable crust results in low earthquake risk; it is located far inland and south from the tropical cyclone tracks; the flatness limits landslip potential; and vegetation is sparse so wildfire risks are few.

The landscape is flat, the horizon is wide and the sky is big. The daytime skies are clear and bright, making them amenable to solar power generation. The night sky is something to behold, with no light pollution, the Southern Cross blazes brightly within the Milky Way as a beacon to the south. The skies present astronomical observing opportunities such as those occurring further west with the Square Kilometre Array.

Population density is very low. Apart from the small towns of Leonora and Laverton, there are remote mine sites with FIFO work camps, scattered pastoral stations (ranches) and several Aboriginal communities.

Although remote, the region is connected to rail, grid power, sealed highways and a gas pipeline. These backbone infrastructure elements place the region favourably for DSP.

The Aboriginal people of Australia are said to be the oldest living culture – they have been 'Caring for Country' for many more than 60,000 years! Their dreamtime stories and lore are deeply found in the unique land and landscape and in the plants and animals.

The oldest rocks of the region were formed 2.7 billion years ago, in the Archaean Era when the planet was very different from today. These old Western Australian rocks were formed in equatorial regions within the Kenorland Supercontinent, which also included fragments of what is now found in Canada, Scandinavia and Southern Africa.

The crust at this time was dominated by granites being pumped into the crust from partial melts of basalts deeper down in the lower crust and mantle and 'greenstones' erupting into basins within this crust.

The greenstones include rocks that, due to the fact that the Earth was hotter in the past, do not form today. They are known as komatiites. They are special rocks as these lavas erupted at more than 1300°C, scorching their way across the land and sea, even eroding channels into the underlying crust. They are also special in the very high amounts of magnesium and high amounts of critical minerals such as nickel, cobalt and platinum group elements that are all essential for decarbonising the world. A number of world-class mines in the region are producing critical minerals from these rocks.

The geological name for the region is the Eastern Goldfields. It is called this because it is rich in gold, as well as other minerals. Gold was first discovered in the late 1800s by hardy prospectors who panned and sunk shafts on likely quartz reefs. Great deposits were found with the Sons of Gwalia mine at Leonora being now more than 1500 metres deep. The area has been an active mining centre for almost 150 years. Most of the gold is found in the greenstone rocks as these host the suitable structures and chemical ingredients to form gold deposits.

Most of the rocks, by area and volume, comprise granites. These rocks can be proud in the landscape and they form beautiful whale back-like edifices. Watching a sunrise or a sunset on top of one of these one can only imagine the Earth events these ancient rocks have 'seen'. Gnamma holes are sites of permanent or semi-permanent water that are commonly located on or adjacent to these granite outcrops (Western Australian Museum, 2022). They are sacred to Aboriginal peoples as they were essential to survival in drought times.

Despite being abundant, the economic resource endowment of the region's granites is limited. Some mines occur in them, but there is potential to consider elements not particularly explored for in the past, such as lithium which is essential for Li-ion batteries.

Great forces of unknown origin heated, bent and buckled the ancient granite-greenstone crust. Such forces drove mineral-rich fluids from down deep upwards into higher levels in the crust where many rich deposits of particularly gold and nickel were formed and preserved. These world-class deposits have been the backbone of the regional economy for more than a century and will be so into the future.

Following these mineralising events, the crust cooled and stabilised – it is known as the Yilgarn Craton (or shield) and is the old hard 'core' of the Australian continent. But it was not to stay fixed.

Kenorland broke up and the fragments (cratons) were distributed across the globe. The Earth tried to break up the Yilgarn Craton many more times. Evidence of great dykes of basalt rock cut deeply across thousands of kilometres is a testament to these enormous forces. The rare-earth deposit at Mt Weld formed about two billion years ago during one of these attempted break-up events. The host rocks are called carbonatites, which were derived directly from the partial melting of the mantle. This is a significant site as it contains rare-earth elements, including the highly desirable heavy elements, which are essential for high-tech manufacturing. Diamonds come from similar depths in the Earth and they have been discovered elsewhere in similar cratonic regions of the globe.

To the East lies a great geological basin – the Officer Basin. It started to form around 825 million years ago during the break-up of the supercontinent Rodinia on the down-warped and faulted eastern edge of the Yilgarn Craton. Over the next 400 million years around eight kilometres of sedimentary rocks were accumulated in this basin, including thick deposits of salt (halite) and glacial outwash sediments developed during several of the global Snowball Earth episodes.

The Australian continent continued to be driven by the great forces of plate tectonics, meandered across the planet and by 300 million years ago Australia found itself centred around the South Pole. Glaciers scoured the surface and in the Laverton region and to the east, thick deposits of sediments were accumulated.

Deep erosion and ongoing weathering followed; the climate changed and Australia as we know it today started to drift north around 34 million years ago to its present geographical position. During this time any remaining mountainous areas were worn down and the landscape was subdued into a vast flat plain lying at an elevation around 300 metres above the present sea level. During periods of higher rainfall, these elevated flat plains were dissected by long river-valley systems a few hundred metres deep. These systems drained southwards into the Southern Ocean and northwards into the Indian Ocean. Increased drying, particularly during the past 5 to 10 million years, resulted in the river valley systems mostly becoming inactive. These systems are presently known as palaeovalleys (ancient valleys) where elongated salt lakes, that like pearls on a string, stretch for across hundreds of kilometres of the landscape.

In the past 60,000 years, as depicted in the Integrated Natural Resources Model (INRM) in Figure 8.10, these lake systems have fluctuated with the climate, shifting between freshwater and salt water. During the more favourable freshwater years, these regions would have provided an environment of abundance for the Aboriginal peoples.

In terms of many things, but soils in particular, Australia is quite unlike the Northern Hemisphere. The glacial and interglacial cycles that sculpted the north are restricted to a few small mountainous areas in the southeast of the continent. In the Northern Hemisphere, many of the soils are young and fertile, having had the old weathered and depleted land surface largely scraped away by ice. Australia's millennia-long weathering and erosion, from the poles to the topics, has created nutritionally poor soils. These soils and the challenges of a hostile climate (arid) mean that the ecosystems have also evolved over the millennia. They have become highly specialised and adapted with a resultant endemic biodiversity almost second to none. What is the potential of pharmaceutical properties of this diversity? What is the risk of extinction? Australia has an unenviable track record of extinctions and damage to ecosystems.

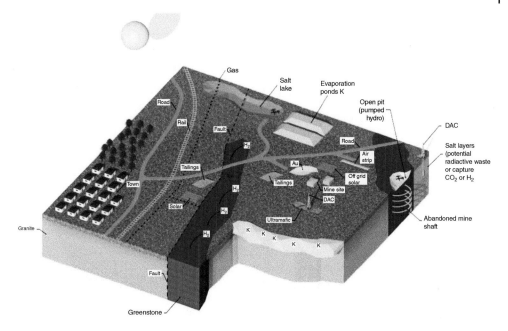

Figure 8.10 An example of an Integrated Natural Resources Model for the Leonora to Laverton region Australia.

In terms of agriculture, the poor soils and insufficient surface water restrict significant cropping. The region only supports sparse ranching of sheep and cattle, although numerous abandoned homesteads attest to the hardships of this style of farming. Aboriginal peoples coveted the sandalwood (*Santalum spicatum*) trees as a source of bush food and medicine and for smoking ceremonies. Colonists began harvesting sandalwood trees to export overseas for incense production, decimating their numbers even in the arid and semi-arid interior. Millions of trees have been exported since the 1840s, pushing the species towards extinction in the wild. Could these trees be replanted and managed sustainably as a future opportunity?

The salt lakes have often been viewed as a challenge and a nuisance for resource development as they conceal the ore-bearing greenstones below. What people did not realise is that some of these salt lakes contain a most valuable resource and in economic quantities. Potash is an essential ingredient for fertiliser. It is particularly important for Australian agriculture as the soils are typically nutrient poor and require supplements. Previously, Australia was not self sufficient in potash and it was one of the few 'critical commodities' the nation needed to import. This is no longer the case, as now a vibrant, DSP-aligned development is occurring on a number of the lakes. The company works with the Aboriginal communities providing ongoing education and skill development and a workforce opportunity. They also use excellent solar resources as part of the processing. Other minerals found in salt lakes are uranium and there is potential – although it is probably low – of lithium contained within the brines.

Water issues are the primary concern when embarking on a DSP initiative, necessitating our focus on securing a sustainable and clean water supply as a foundational step. We have described the arid region as largely devoid of significant permanent surface water. There is, however, water below the surface – as groundwater. Groundwater rarely exists as underground lakes or rivers. Rather the water occurs in the pores between the grains that make up a rock or in fractures and gaps in the rock. The most plentiful groundwater is found in the palaeovalleys that stretch widely across the landscape, the water can be both salty and fresh in composition. Further east in the Officer Basin are vast untapped groundwater reservoirs. Questions need to be asked about the recharge rates and thus sustainability of developing and using these waters as most are ancient fossil water accumulated in millennia past when rainfall was more abundant than today.

The very high magnesium content of some of the greenstone rocks weather to a white-coloured rock. This natural weathering involves forming new carbonates such as $MgCO_3$ with the CO_2 derived directly from the atmosphere. Opportunities for sequestering atmospheric or point-source CO_2 could be developed within the region, with extra consideration given to existing mine tailing sites where large volumes of crushed material lie wasted and unused but could be transformed into carbon sinks. The development of direct-air capture (DAC) units powered by the sun is progressing. Marrying arrays of DAC units and suitable greenstone geology could be an opportunity for DSP.

Hydrogen is the lightest element and is notoriously difficult to capture and control. Hydrogen, if managed, could be a zero carbon fuel source. Hydrogen is known to be degassing from the breakdown of minerals in greenstone rocks in some mines. Mapping and measuring the hydrogen emissions would enable an evaluation of the potential for the region. Burning hydrogen creates water and energy, which can power local sites or be diluted into existing gas pipelines and delivered in a mix with natural gas (methane) into larger processing centres. Opportunities for hydrogen storage could be sought in the salt deposits and caverns of the Officer Basin to the east.

The mining and ore recovery processes at most mine sites are engineered to focus on single or selected elements. The result is mine tailing dams containing a treasure trove of elements that are now known to be critical for modern technology, including those required to decarbonise. The region has a number of sites that could be amenable to supply such elements, providing another possible DSP opportunity.

Depleted mine sites and open pits and underground shafts are scattered across the region. Many of the pits are more than one kilometre long and hundreds of metres deep and filled with water of variable quality and chemistry. These sites are unused, but opportunities might exist for developing pumped hydrostorage in combination with solar, either in the open flooded pits or in combination with the pits and their underground workings. Testing the water quality could ascertain that some pits – now 'lakes' could be developed for aquaculture or even tourism.

Almost all greenstone belts have been explored to some extent, but there is enormous potential for the discovery of new deposits of minerals hidden under the basins, salt lakes and sands that blanket the region. The technology for seeing into the Earth is highly advanced, with Australia being the best in the world at using it. Opportunities for a global showcasing or centre of excellence for successful undercover exploration and discovery could be investigated.

Millennia of traditional knowledge is vested with the Aboriginal peoples. DSP opportunities could be examined for a range of cases including: bush medicine, remediation and recovery of past or future development and eco-tourism, amongst others. The townsfolk and traditional owners are proud of their heritage and they are willing to look to a positive mutually beneficial future.

Northern Ontario's 'Ring of Fire' – A Nature Positive Solution

Alignment to UN SDGs 1, 2, 3, 4, 6, 7, 8, 11, 12, 13, 14, 15, 17

Summary

The 'Ring of Fire' in Northern Ontario, shown in Figure 8.11, represents a globally signifi-cant ecosystem, offering both opportunities and challenges. Covering one of the world's largest wetlands, this remote rural region is surrounded by Indigenous communities, neces-sitating a balanced approach considering environmental, social, cultural and economic aspects.

Canadian peatlands encompass nearly half of the earth's wetland cover (Harris, et al., 2021). Occupying 13% of Canada's vast land area (Tarnocai, Kettles, & Lacelle, 2011), with a particularly significant presence in northern Ontario. They play a pivotal role in retaining water, sequestering around 60% of Canada's organic carbon in soils, mitigating flood, purifying water and providing vital wildlife habitats.

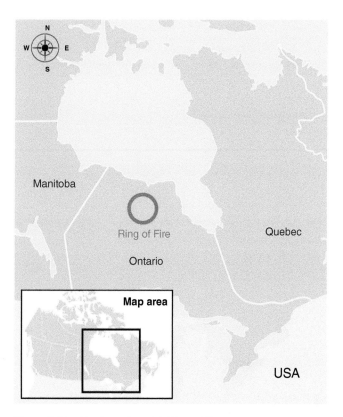

Figure 8.11 Northern Ontario's Ring of Fire.

While the region boasts abundant mineral resources, notably chromite deposits and sig-nificant reserves of nickel, copper, gold, platinum-group elements and zinc (Ontario, 2022), its peatlands are even more valuable, acting as natural carbon sinks.

With limited access to nearby cities such as Thunder Bay and Timmins, the region balances natural wealth and ecological responsibility. Applying DSP principles, a nature positive solution focused on peatland regeneration emerges as a pathway to sustainable development.

Current Situation

The Ring of Fire's current situation combines ecological wealth, mineral potential and cultural diversity and infrastructural challenges.

1) **Environmental Significance**
   - Peatlands are vital for carbon sequestration and water management. Peat soil can hold up to 30 times its weight in water.
   - The region's wetlands play a pivotal role in global warming mitigation.
2) **Geographical and Mineral Profile**
   - The area covers $5120\,km^2$ in the James Bay Lowlands.
   - Beneath the peatlands are valuable deposits, including chromite, copper, nickel, platinum, vanadium and gold.
3) **Indigenous Communities**
   - Ontario's far north houses 40% of the Indigenious populations.
   - Indigenous communities consider themselves stewards of the land, aspire to prosperity and responsible investment and some oppose development.
4) **Infrastructure and Resources**
   - Limited transportation infrastructure affects accessibility.
   - High energy costs pose challenges.
   - A shortage of skilled labour.

Untapped Potential

The central challenge lies in balancing wetland ecosystem preservation with responsible extraction in collaboration with local communities.

Figure 8.12 shows a healthy peatland ecosystem working as a carbon sink by accumulating dissolved organic carbon (DOC). The waterlogged environment slows down plant decomposition, leading to peat formation.

Potential Scenario in the Ring of Fire

1) **Ecosystem Preservation**
   - Maintain the wetland's health for ecological balance.
   - Enhance water management for biodiversity production.
   - Prevent wildfires by keeping peatlands healthy.
2) **Carbon Storage and Climate Action**
   - Peatlands store more carbon than any other terrestrial ecosystem globally (Andersen, 2017). They store three times more carbon than a boreal or tropical forest for the same surface area.
   - Restoration and conservation of peatlands contribute to achieving net zero greenhouse gas (GHG) emissions.

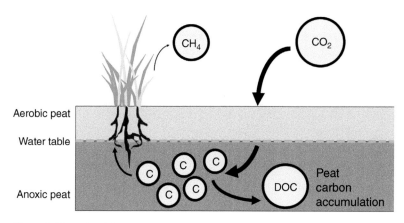

Figure 8.12 A healthy peatland.

3) **Innovative Conservation**
   - Pioneer new conservation methods and technologies that could have a global impact.
   - Sustainably restore damaged peatlands. Tanneberger highlights that rewetted peatlands offer potential for carbon sequestration, improved water quality and sustainable biomass production (Tanneberger, 2021).
4) **Indigenous Wisdom and Storytelling**
   - Leverage the rich cultural heritage of the indigenous communities for conservation efforts.
   - Use storytelling as a powerful tool for environmental stewardship.
5) **Collaborative Development**
   - Develop eco friendly infrastructure.
   - Advance responsible mining and environmental monitoring technologies.
   - Create resilience through conservation-based enterprises.

There is a potential to transform the 'Ring of Fire' into a sustainable development model, setting global benchmarks in sustainable practices.

Challenges and Risks

The central challenge for the region centres around the following question: 'How can we leverage the abundant natural resources of this remote location while preserving the fragile environment and respecting the culture of Indigenous communities?'

1) **Carbon Emission Concerns**
   - The challenge, therefore, is how do you avoid damaging the peatlands. Damaged peatlands release stored carbon, undermining global efforts against reducing global warming.
   - Disturbed peatlands emit approximately two billion tonnes of CO_2 annually, accounting for roughly 5% of global GHG emissions (United Nations Environment Programme, 2023). Emissions from peatlands are expected to rise sharply. As the threat of climate change has grown more severe, researchers and governments have identified peatlands as ideal targets for stopping emissions.
2) **Conservation and Restoration Initiatives**
   - Scotland stands out as a leader in habitat restoration, focusing on over 20% of its land covered by peatlands (World Wildlife Fund, 2023).

- Globally, wetlands, covering only 3% of the Earth's surface, store 30% of land-based carbon (Economist Impact, 2022). Methods of conservation and restoration include revegetation of the disturbed site and rewetting the peat surface (Girard, Lavoie, & Theriault, 2002; Smolders, Tomassen, Lamers, Lomans, & Roelofs, 2002).
- Meeting the Paris Agreement's 1.5°C target requires not only protecting existing wetlands but also restoring 50% of all lost wetlands by 2030 (Ramsar Convention on Wetlands, n.d.).

3) **Biodiversity Loss**
- Degraded peatlands can lead to the loss of unique and sometimes endangered species.

4) **Water Management Issues**
- Damaged peatlands affect water retention and purification, affecting both humans and wildlife.

5) **Increased Fire Risks**
- Damaged peatlands become susceptible to fires, which can have devastating effects on the surrounding environment and communities.

6) **Predictive Modelling Gaps**
- Predicting the impacts of changing water supply and climate on peatland ecosystems is challenging.

Understanding and addressing these challenges are vital for ensuring the sustainability and health of these unique ecosystems.

Potential Future Offered by DSP

The 'Ring of Fire' with its vast natural wealth and biodiversity, presents a unique opportunity for transformation through the application of DSP principles. The challenge is to balance resource extraction with the crucial preservation of region's peatland ecosystem, promoting sustainability and ecological stewardship.

1) **Peatland Regeneration**
- Understanding peatland hydrology is essential to stabilise, regenerate and maintain these vital ecosystems. Line Rochefort developed a new approach to restore North American peatlands (Rochefort, Quinty, Campeau, Johnson, & Malterer, 2003).
- Enhancing CO_2 capture through peatland formation while reducing methane and nitrous oxide emissions.
- Implementing advanced water management combined with productive wetland culture to accelerate moss reestablishment.
- Exploring geoengineering solutions to mitigate GHG emissions.

2) **Strategic Approaches**
- Collaborating with global institutions to gain critical knowledge for peatland regeneration and developing tailored regeneration techniques. 'It's so easy to break an ecosystem and it's so hard to bring it back', says Andersen (Andersen, 2017). Andersen's research underscores the importance of collaboration to determine how to best manage the peatlands for carbon storage. Extensive research has been conducted in the United Kingdom as well as in Canada on peatland restoration. By collaborating with such institutions and developing regeneration practices, this region can enhance its understanding and management of these complex ecosystems.

- Focusing on minimum impact mining methods, such as underground and remote operations to minimise surface disturbance, setting new standards in sustainable mining.
- Leveraging the power of narrative to educate and inspire stewardship of the peatland ecosystem.

3) **Research and Innovation**
   - Recognising the specific hydrological functions (storage, transmission and runoff) of wetlands and their connectivity within the landscape (Goodbrand, Westbrook, & van der Kamp, 2019). Wetlands are specific components of groundwater flow systems and their connectivity is affected by their position in the landscape, the aquifer geology and the climate setting (Bourgault, Larocque, & Garneau, 2019).
   - Addressing climate projections for a warmer and wetter climate, impacting groundwater recharge and water table depth and consequently impacting the influx of groundwater to peatlands.
   - Restoring natural hydrology by selecting suitable locations for recovery, allowing systems to rebound naturally.

4) **Global Leadership Opportunity**
   - Canada, with its intact northern peatlands, can lead in sustainable peatland practices for carbon capture.
   - Establishing a research centre dedicated to developing the know-how in carbon capture efficiency and water management for peatland regeneration, fostering the creation of related enterprises and regional development efforts.
   - Investigating practical methods to convert degraded peatlands into robust GHG removal systems.

5) **Multidisciplinary Skill Development**
   - Equipping the local workforce with necessary skills and tools for peatland transformation and sustainable management.
   - Engaging ecologists, biochemists, hydrologists, restoration scientists, bioenergy experts and social scientists to collaborate effectively in peatland projects for peatland transformation into GHG removal systems:
     - Ecologists and biochemists to understand soil processes.
     - Hydrologists to understand peatland formation and function in response to regeneration practices.
     - Bioenergy scientists to identify potential long term carbon storage mechanisms.
     - Social scientists to understand the drivers of peatland degradation and identify socially acceptable and economically viable alternative management strategies.
   - Tailoring education to these areas will be crucial to equip the local workforce with the necessary skills and tools to support and sustain this effort.

6) **The Role of Storytelling**
   - Creating a Culture of Stewardship: Promoting a narrative where everyone plays a role in protecting the peatland, leading to the preservation of the entire ecosystem.
   - Creating narratives for eco-tourism integrated with peatland regeneration.
   - Blending environmental conservation with sustainable economic opportunities for local communities.
   - Crafting a new narrative for the region rooted in respect for nature, collaboration and a shared commitment to a sustainable and prosperous future.

7) Comprehensive Engagement and Collaboration
- Aligning interests of indigenous communities, mining investments, educational initiatives and government policies for socio-economic expansion.
- Developing a national strategy for peatland regeneration to enhance understanding and management of these complex ecosystems.

This multifaceted approach holds the potential to transform the 'Ring of Fire' into a model of sustainable development. It seeks to balance ecological conservation with responsible resource extraction, aiming to create a resilient and thriving region.

Figure 8.13 illustrates a vision for the 'Ring of Fire' that prioritises nature positive solutions. At its core lies the strategic focus on peatland regeneration serving as the foundation for transitioning into a knowledge-based economy. This know-how is centred on carbon sink solutions, harnessing the region's important peatland resources and their critical role in global climate regulation. Mining activities serve as a catalyst in realising this transformative journey.

Conclusion

The 'Ring of Fire' presents an opportunity to redefine our relationship with nature and serve as a global model for balanced development. The main challenge is to maintain the environmental benefits of the peatlands including retaining the carbon sink features.

Figure 8.13 'Ring of Fire' – Prioritising nature positive carbon sink solutions.

Peatland regeneration is the key to balancing natural wealth with ecological responsibility, prioritising sustainability and serving as a global model for balanced development.

To achieve this balance the priority must be peatland regeneration, this benefits all and the mining investor becomes the catalyst to turn this region into a hub for the knowledge of the peatland hydrology for peatland regeneration to impact the climate locally and globally.

Collaboration among key parties can turn this region into a hub for peatland knowledge and climate impact.

It offers a model for sustainable development worldwide, symbolising hope, resilience and commitment to a more sustainable world.

We have only taken the first steps in the DSP approach by discovering what possibilities and challenges lie ahead. We would need to engage with the community and key parties in the region to co-create this future.

Northwest British Columbia – Low Carbon Solutions Hub

Alignment to UN SDGs 1, 2, 3, 4, 6, 7, 8, 11, 12, 13, 14, 15, 17

Summary

In Northwest British Columbia (NWBC), the stunning natural environment is a fundamental element of the ecosystem that must be central to regional development. The area's rich cultural heritage and the presence of Indigenous communities with deep rooted traditions necessitate their integral involvement in any development process. The DSP approach in this region requires a deep respect for the environment and a commitment to intimate collaboration with local communities. This ensures that development solutions are holistic, honouring both the ecological integrity and the cultural significance of the region.

In this region, DSP was applied to accelerate mining development while simultaneously upholding environment protection and respecting the rich cultures of the Indigenous communities. This approach ensured that the advancement of the mining sector contributed positively to the region, aligning with sustainable practices and cultural sensitivity.

Leveraging its abundant natural resources supported with good infrastructure, the initiative focuses on a collaborative mining economy emphasising shared resources and remote operation of mines. The acquired expertise in remote sensing knowledge could be expanded to wildlife monitoring and avalanche detection. This region with its robust hydropower supply, is uniquely positioned to evolve into a low carbon solutions hub. By integrating its strength in hydropower with innovative approaches in remote sensing, the region could set a benchmark in environmentally sustainable practices.

The solution therefore is:

- To develop a Remote Collaborative Mining Ecosystem.
- The knowledge economy will centre around remote sensing technologies and low carbon energy solutions, which are core to the success of sustainable remote mining and can also be leveraged in global climate disaster prevention and energy optimisation.

Current State

NWBC, as shown in Figure 8.14, is in a semi-remote region. This region hosts abundant gold, silver, nickel and copper deposits and forestry and agricultural industries. There is also established infrastructure including hydropower, the nearby port of Stewart and Prince Rupert,

Highway 37 running north-south through the area and rail lines connecting Prince Rupert, Prince George and Smithers. The 'Golden Triangle' received 10% of Canada's exploration spending in 2020 (The Interior News, 2021), highlighting its importance in the gold and copper rush of British Columbia in the past. The region has good existing infrastructure, but the mountainous terrain and fragile environment make access and mining challenging. Additionally, several Indigenous communities reside in the area, necessitating collaborative efforts for sustainable development.

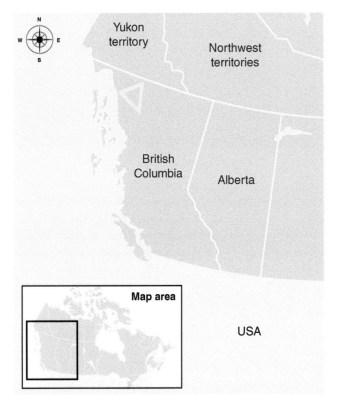

Figure 8.14 Location of NWBC's 'Golden Triangle'.

Untapped Opportunities

The DSP preliminary assessment reveals several untapped opportunities:

- Entrepreneurial Indigenous people with rich cultures and values can lead to development of sustainable enterprises that incorporate traditional knowledge and practices.
- Abundant, clean and affordable hydropower is a significant advantage for the region to enable sustainable energy generation and to diversify the local economy.
- Infrastructure including ports, rail and roads. While the existing infrastructure provides a solid foundation, there are opportunities to expand and enhance transportation networks, digital infrastructure and waste treatment solutions.
- A region rich in essential minerals offers an opportunity for responsible and sustainable mining while protecting the environment, such as with remote mining.
- Proximity to the forestry, agriculture and maritime resources facilitates international trade and opens opportunities for these industries.

- Highly educated individuals can contribute to capacity building and the development of new technologies and practices that can help address the unique challenges faced by the region, such as environmental preservation and sustainable mining.
- Proximity to a volcanic region makes NWBC one of the most seismically active regions in Canada which provides good geothermal potential for renewable power generation, as shown in Figure 8.15.

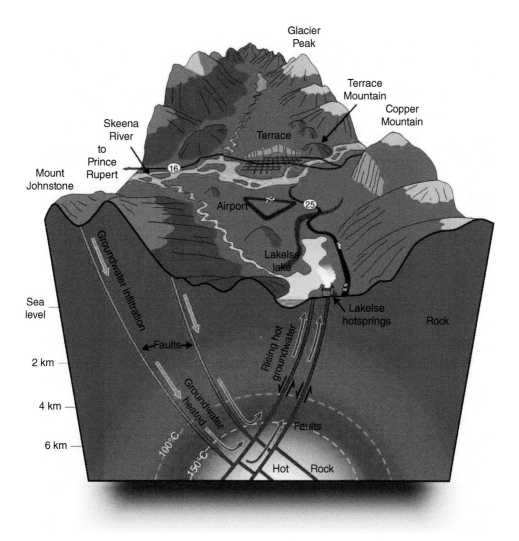

Figure 8.15 Schematic of Lakelse Spring, British Columbia, illustrating integrated geological faults, groundwater and volcanology in geothermal potential. *Source:* Grasby, et al. (2012) / with permission of Her Majesty the Queen in Right of Canada.

Challenges and Risks

The central challenge for the region centres around the following question: 'How can we leverage the abundant natural resources of this semi-remote location while preserving the fragile environment and respecting the culture of Indigenous communities?'

Traditional mining operations risk adversely impacting the surrounding environment. Mine waste, tailings and water pollution can irreversibly damage the environment, affecting ecological life and human populations. As such, innovative solutions and collaboration efforts to develop sustainable mining operations while protecting the environment are required.

The risk of changes to the environment due to mining, compounded with climate change, can lead to an increase in natural disasters. NWBC is already subjected to seismic activity, tsunamis, precipitation extremes and avalanches, all of which disrupt local businesses and damage infrastructure. Solutions for early warning and detection are critical for the communities in the area.

Lastly, mining operations can displace local communities and cultures. Such operations traditionally see an influx of workers and development of local infrastructure and community to support the operation, followed by a mass exodus of investment when the mine ends, hence leaving the local economy and the community abandoned and economically adversely impacted.

Since NWBC is also home to Indigenous communities, their involvement in future planning must be respected. The Indigenous communities are entrepreneurial and can provide insights as to how best to work together with the natural gifts of the area. It is essential to work in collaboration with them to ensure their cultures are preserved and sustainable economic development is assured beyond the lifespan of the mine.

All these risks are interconnected and require a holistic approach to create a resilient region.

Potential Future Offered by DSP

The DSP approach focused on identifying avenues to accelerate mining development in this region and to extend the life of these mines while protecting the environment and the cultures of the Indigenous communities.

One potential solution is to set up mining industry with shared infrastructure, equipment and maintenance and use new sensor technologies for remote operations to make these new mining economies economically viable while protecting the environment and the cultures of local communities. This is now being known as a 'collaborative economy'. This sustainable and prosperous NWBC, illustrated in Figure 8.16, includes the following main opportunities for development:

- **Collaborative Mining as a Catalyst for Regional Prosperity**
  Instead of establishing extensive infrastructure near the mines, the community can combine efforts and investments and leverage the existing infrastructure in the region to form a collaborative mining industry. For example, shared pipelines to transport concentrate or fuel and fibre optics to have good communications and reap the benefits of the digital age. By pooling investment and sharing resources in a collaborative manner, the growth of the mining industry can be accelerated. Indigenous communities must play a key role in this collaboration scheme.
- **Building a Knowledge Economy Around Technological Enablers for Collaborative Remote Mining**
  - **Sensor Technology**
    The implementation of remote operations will cultivate expertise in sensor development; a skill set with broad applications beyond mining. Sensor know-how can be applied to smart grids, optimising energy consumption, monitoring biodiversity and safeguarding the environment. As the region hones its capabilities in sensor development, it could emerge as a leader in acoustic technologies; a field of importance in a

region where avalanches pose a significant risk. Acoustic sensors can provide early warning signals for avalanches, monitor the health of manufacturing equipment and find applications in various other domains.

Advanced remote sensing technologies, such as acoustic sensing, which characterises structure and dynamics of terrestrial and aquatic systems, can play a huge role in not only enabling remote mining, but also in the adjacent energy sector. Acoustic sensing presents immense value for the energy value chain, from online and real-time asset management of green energy generation and energy storage systems, to improving building energy efficiency by monitoring occupancy levels. The same technology, along with adjacent sensing technologies such as acoustic fibre optic sensing, satellite remote sensing, radar sensing and infrasound sensing can be transferred to measure sound in air, water and solids – that is, applications in early warning and detection systems for avalanche and flood, all of which have played a detrimental role in the BC area in recent years as a result of climate change. New education centres focusing on sensor technology, sensor data processing, control systems and digital transformation, can be opened in the hubs, as satellite centres to larger colleges and universities in BC, attracting external skilled professionals while enabling higher education for local communities. The new education centres can also serve to export remote sensing technology know-how globally, which will become increasingly critical as we continue to improve sustainability in our natural resources industries while combating climate change challenges and unprecedent natural disasters.

The mastery of acoustic technologies not only addresses the immediate challenges faced by the region but also holds the promise of unlocking new opportunities. By developing cutting-edge solutions for avalanche warning systems, NWBC can contribute to global efforts in disaster prevention and mitigation. Furthermore, expertise in equipment health monitoring can drive innovation in manufacturing, leading to more efficient and sustainable operations.

– **Low Carbon Energy and Water Treatment Technology**
With its wealth of hydropower, the region is uniquely positioned to lead in sustainable power generation, which is integral to supporting remote mining operations. This pivotal role underscores the importance of proficient water management, ensuring the responsible and efficient utilisation of hydropower. By optimising water resources, NWBC can fuel remote mining operations with clean, renewable energy, drastically reducing its carbon footprint. BC colleges and universities offer numerous programs in mining (University of British Columbia's Institute of Mining Engineering, British Columbia Institute of Technology's Mineral Exploration and Mining Technology) and in ecology (British Columbia Institute of Technology's Ecological Restoration, University of Northern British Columbia's Forest Ecology and Management), setting up a solid foundation for growth of NWBC into a knowledge centre for new sustainable mining waste management and clean energy generation and supporting programs such as information technology, data storage, automation and ecological protection.

• **Exporting the Knowledge Economy – Building Communities that Survive Beyond the Lives of the Mines**
The knowledge economy around remote collaborative mining is translatable to other industries, which ensures resilience and sustainability in the NWBC communities. As NWBC expands beyond mining, the same expertise can be leveraged to diversify its

economy, creating new opportunities in sectors such as smart grid development, energy optimisation and environmental monitoring. This diversification will ensure the region's long term prosperity, reducing its dependence on mining and fostering a more resilient and sustainable region.

By working with local Indigenous communities and presenting the opportunity to drive this effort jointly, we can identify synergistic industries that help with improving their quality of life while preserving their culture and environment. The region's forest and land environment and proximity to marine environment equate to opportunities in forestry and agricultural sectors. Identification of high value forestry and agricultural niches, such as value added food product industries around regional wild berries, seaweed and fish products as well as high-tech greenhouse technology, are possibilities. Coincidentally, technology and skills associated with selective fishing are transferable and applicable to selective mining technologies.

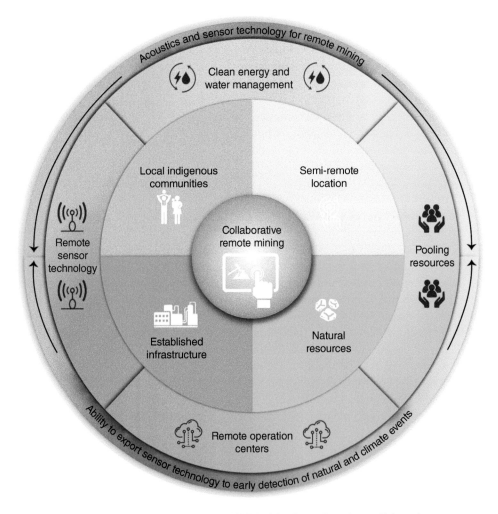

Figure 8.16 Possible sustainably prosperous NWBC of the future based on collaborative remote mining.

By applying the principles of DSP in planning a diversified economy on top of a mining economy, a sustainable and long-living mining economy can flourish in NWBC while protecting the surrounding environment in collaboration with Indigenous communities. A sustainably prosperous NWBC could be built upon a knowledge industry centred around remote operations, including developing specialised skills and knowledge to address the industries and opportunities of the local community and exporting and transferring the knowledge globally into adjacent industries.

Conclusion

The DSP solution envisages a future for NWBC, where the region leverages its natural resources and develops expertise in remote operations, sensor development, acoustic technologies, low carbon energy production and water management. This multifaceted approach will facilitate sustainable mining operations while protecting the surrounding environment in collaboration with neighbouring Indigenous communities. In addition, the know-how developed for sustainable remote mining can also contribute to global efforts in disaster prevention, energy optimisation and environmental preservation. By embracing this vision, NWBC can transform itself into a global centre of excellence for remote collaborative mining, creating a legacy of prosperity and sustainability for its inhabitants and the world. This transformation will redefine the region's role in the global landscape, transitioning from a mining-dependent economy to a diversified, knowledge-driven hub, contributing to a more sustainable and resilient future.

We have only taken the first steps in the DSP approach by discovering what possibilities lie ahead and what the design basis of the future ecosystem could look like. We would need to engage with the community and key parties in the region to co-create this future.

Quebec Region, Canada – An Integrated Green Energy System

Alignment to UN SDGs 1, 2, 3, 4, 6, 7, 8, 11, 12, 13, 14, 15, 17

Summary

Quebec, Canada, stands as a region celebrated for its natural beauty and abundant resources (Figure 8.17). The province boasts a strong economy that is closely tied to mining and metals powered by abundant hydropower. However, a significant challenge lies ahead in diversifying the economy, as clean power capacity approaches its limits in the next decade due to decarbonisation goals. Additionally, the region confronts the global challenge of climate change, necessitating a reduction in carbon footprint and waste generation and transition to more renewable energy sources. There is also the need to address the ageing population and workforce by attracting younger professionals and ensuring knowledge transfer.

The DSP approach proposes transforming the economy into an 'Integrated Clean Energy Ecosystem'. This involves harnessing natural resources, existing infrastructure and a workforce to transition towards 'green' electricity and fuels while also attracting young professionals and fostering the knowledge economy, all while respecting the natural environment. This ecosystem optimally matches and controls power sources and users and optimises water reservoir levels throughout the seasons by complementing hydropower with renewable energy and energy storage. This not only addresses Quebec's current challenges but also serves as a global model to transition to renewable energy.

- The region's ultimate need is to diversify its economy.
- The challenge is to achieve decarbonisation by 2050 which requires doubling the existing energy production capacity by Hydro-Quebec.
- There is a need to develop an Integrated Clean Energy Ecosystem.
- The knowledge economy will centre around the efficient management of this complex ecosystem and the ability to provide a stable supply to the grid, thus ensuring resilience.

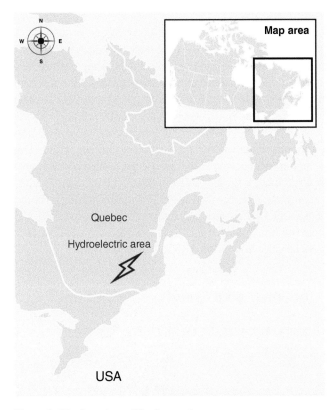

Figure 8.17   Overview of Quebec region.

Current State

The Quebec region ranks among the world's largest producers of hydropower, with a thriving economy encompassing iron ore, copper, gold and aluminium refining. It also enjoys a solid foundation of well connected infrastructure and a highly skilled local workforce. Nevertheless, it faces several challenges, including an impending clean power shortage, high carbon emissions from industry and transportation, an ageing population and the need to increase the availability of renewable energy sources.

The region is home to indigenous communities whose activities include seal hunting, salmon fishing and the protection of the magnificent surrounding natural environment. During the nineteenth century, colonial settlers exploited the abundant forest and access to seaways, establishing lumberyards and sawmills. The early twentieth century witnessed the growth of mineral extraction and processing industries.

Quebec's economy primarily relies on minerals and metals, with contributions from chemicals, agribusiness and tourism, all capitalising on the region's natural endowments. However, there is a need to reduce the carbon footprint, waste generation, transition to renewable energy and attract a younger population.

The region benefits from:

1) Established hydropower providing clean, low cost and on demand energy.
2) Robust infrastructure network including airports, railways, roads and deep sea ports.
3) Connection to major Canadian, US, and European markets.
4) A skilled workforce offering technical expertise at competitive costs, complemented by research parks.

Market Needs

To achieve decarbonisation in a developed region, the following areas need to be addressed:

- **Energy Production and Management**
  - Transition to renewable energy sources including hydropower, wind, solar and hydrogen.
  - Development of charging networks for electric vehicles.
  - Promotion of self sufficient energy homes incorporating the latest green technologies.
  - Expansion of biomass and bioenergy sectors
  - Implementation of waste valorisation strategies to convert waste into energy and other valuable products such as low carbon fuels.
- **Agricultural Sector Transformation**
  - Implementation of sustainable agricultural practices.
  - Development of eco friendly fertilisers and efficient fish farms.
- **Infrastructure Evolution**
  - Rebuilding and reconfiguring urban and rural layouts to support sustainable living.
  - Increasing public transportation options to reduce reliance on personal vehicles.
  - Fostering localised community developments with walk to amenities, reducing the need for mega stores.
  - Investing in high speed transport links and IT infrastructure for better connectivity with major cities.
- **Demographic Adaptations**
  - Developing sustainable care solutions to support an ageing population.
  - Creating initiatives to attract and retain younger demographics, promoting regional expansion and dynamism.
  - Establishing programs for knowledge transfer from older to younger generations, preserving and fostering innovation.

Untapped Opportunities

Quebec has the potential to transform its economy into a renewable power base by leveraging its existing infrastructure, skilled workforce and natural resources. Opportunities include transitioning to electric, hydrogen or hybrid transportation; developing a green hydrogen production industry; producing 'green' metals, chemicals and agribusiness industries; and attracting young professionals to the region.

The availability of large water reservoirs and existing workforce in mineral processing and metal refining are seeds for starting a green hydrogen production industry. This transformation can support the growth of a knowledge economy based on efficiently managing complex green energy ecosystems, refreshing the workforce with younger talent and creating a synergistic economic ecosystem that addresses future societal concerns while respecting the natural environment.

Quebec benefits from a solid foundation of excellent and well connected infrastructure, with access to renewable energy and a highly skilled local workforce. The region is interconnected with cities like Quebec City and Montreal, by rail and highways and internationally by air and deep seaports.

Historical economic growth has led to the establishment of metals, chemicals and agribusiness plants, including prominent aluminium smelters, refineries and cast houses, among other metals exploration and refining operations and the forest products industry. Supporting community infrastructure, such as hospitals, colleges and universities has been established. The local geography and climate make hydropower and wind power easily achievable.

The current economy in aluminium, energy, agribusiness, metals and metallurgy, medical research, forestry and tourism has fostered a skilled and diverse population, with unique expertise that ranges from manufacturing to health care to high-tech.

Untapped potential in the region could be outlined as follows:

- Beyond established hydropower, there is significant untapped potential for wind and solar power development throughout Quebec but will require massive investment and additional transmission lines.
- Leveraging Quebec's large water reservoirs and the possibility for variable hydrogen production that follows the variation of renewable power to enable more renewable power integration on Quebec's electrical system.
- The existing infrastructure network has the capacity for further use, potentially facilitating greater clean and green industry development.
- The presence of a skilled technical workforce could be directed towards innovative industries, particularly in renewable energy and technology sectors.
- The experience of an ageing population could be harnessed in mentorship roles, providing wisdom and stability in a transitioning economy.
- Research centres and industrial parks offer untapped potential for startups and established enterprises to innovate, particularly in green technology and sustainable practices.
- The strategic location with access to major Canadian and US cities presents untapped opportunities for expanding into new markets, especially for the knowledge economy for green products and services.

Challenges and Risks

One of the biggest challenges for our global society is climate change. Quebec emits 79 million tonnes of CO_2 equivalents per year – equivalent to 29 gigalitres of diesel, resulting from fuel consumption in industry and transportation. Quebec's emissions per capita are the lowest in Canada at 8.9 tonnes of CO_2 equivalents – 50% below the Canadian average of 17.7 tonnes per capita but there are opportunities to significantly reduce that figure. The largest emitting sectors in Quebec are transportation at 39% of emissions, industry including manufacturing at 26% and buildings (residential and commercial) at 12% (Canada Energy Regulator, 2023).

The transition to a renewable power ecosystem involves significant challenges, including technological feasibility, infrastructure development, financial investment and environmental impact. Replacing diesel, gasoline and heavy fuel oil with green alternatives requires a supporting fuelling infrastructure and may pose challenges in industrial production. Additionally, there is a need to manage the environmental impact of this transition, ensuring the sustainability of the surrounding natural environment.

The central challenge for the region is the need to achieve decarbonisation by 2050, amidst a reliance on hydropower that while clean, falls short of fully supplanting fossil fuels. With the timeline for constructing new hydropower plants stretching up to two decades, there is an urgent need to identify and implement alternative solutions that can accelerate the transition to a low carbon future within the required timeframe.

Potential Future Offered by DSP

The DSP approach identified a unique way to transform the region's economy and skillsets into an 'Integrated Clean Energy Ecosystem' to generate a stable supply of electricity to the Grid (refer to Figure 8.18). This involves managing and integrating a complex industrial ecosystem based on green energy, transitioning to electric, hydrogen or hybrid transportation, developing a green hydrogen production industry and supporting the expansion of 'green' metals, chemicals and agribusiness industries. By-products from the metals and chemicals industries can flow to local users, such as oxygen for healthcare and potassium salts for agriculture. This will lead to the creation of a synergistic economic ecosystem that addresses future societal concerns while respecting the natural environment.

Collaboration between government, community and industry is crucial to develop this vision into reality. The know-how in implementing this 'Clean Energy Ecosystem' can be perfected in Quebec and then exported globally.

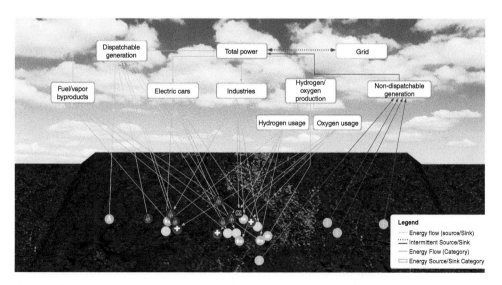

Figure 8.18 Possible Integrated Clean Energy Ecosystem of a Sustainable and Resilient Quebec.

The key elements of the transition are:

- A foundation for managing and integrating a complex industrial ecosystem based on green energy.
- Replacing diesel, gasoline and heavy fuel oil used in industry and transportation with carbon free alternatives.
- Transportation decarbonisation by switching to electric, hydrogen or hybrid transportation, with supporting fuelling infrastructure.
- Switching industrial production to electric process or hydrogen fuel based processes.
- The region is amenable to large scale wind power production but also solar. The access to renewable power, large water reservoirs and the existing workforce in metal processing are seeds for the green hydrogen production industry based on electrolysis.
- With this solid bedrock of green electricity and fuels, 'green' metals and chemicals production and agribusiness could expand.
- The metal and chemical industries can support an energy storage and battery industry, which would allow more renewable power storage.
- By-products from the metals and chemicals industries can flow to local users. For example, oxygen in health care and potassium salts, such as potash as fertilisers.
- This synergistic economic ecosystem along with quality infrastructure and the surrounding natural environment would attract talented young professionals to the region.
- The knowledge economy through colleges, universities and industrial communities can expand while augmenting and replenishing the ageing workforce.
- A green Energy Hub to generate a stable supply of electricity to the grid. This could generate more electricity than the region needs now and generate opportunities for new sustainable enterprises to use that excess energy and export that excess.
- The knowledge economy will centre around the efficient management of this complex system. This will enable providing a stable supply of electricity to the grid, thereby ensuring resilience.

Conclusion

A sustainably prosperous Quebec can be achieved by leveraging the natural strengths of the region to create an interconnected resilient ecosystem that addresses future societal concerns.

The DSP approach provides a roadmap for transforming the existing industries and workforce into a green energy system to generate a stable supply of electricity to the grid, where power sources and sinks are optimally managed and renewable power generation, hydrogen generation and energy storage systems complement existing hydropower generation.

This transformation will attract talented young professionals, support the growth of the knowledge economy and ensure the sustainability of the community and economy while respecting the blessings of the natural environment. This vision can be developed into reality by engaging with community and industrial key parties to secure support and financial investment. Ultimately, the unique know-how developed in implementing this 'Clean Energy Ecosystem' can be perfected in Quebec and exported globally.

We have only taken the first steps in the DSP approach by discovering what possibilities lie ahead and what the design basis of the future ecosystem could look like. We would need to engage with the community and key parties in the region to co-create this future. Figure 8.19 illustrates a path of high impact for Quebec, serving as a global blueprint for renewable energy transition.

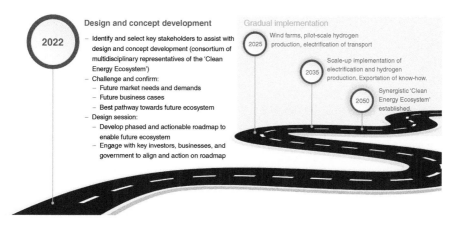

Figure 8.19 Possible path towards 'Clean Energy Ecosystem' for a sustainable and resilient Quebec.

Peru's Sechura Desert, Peru – A Circular Industrial Cluster

Alignment to UN SDGs 1, 2, 3, 4, 6, 7, 8, 11, 12, 13, 14, 15, 17

Sechura Desert, as shown in Figure 8.20, is a coastal desert located in northern Peru close to the border with Ecuador and between the Pacific Ocean and Andes Mountains.

Figure 8.20 Location of Sechura Desert in Peru and indicative solar irradiance, where red is high and green is low.

The region boasts a unique wealth of resources:

- Phosphate and potash
- Copper ore deposits (subsequently sulphuric acid as a by-product if the copper ores are refined)
- Natural gas
- Water from the sea and mountains
- One of the richest sources of fish in the world
- Strong winds and abundant sunshine that could be harnessed for electricity.

Currently, natural resources in the region consist of phosphate, potash, diatomite, calcareous gypsum and natural gas, with projects in metals (copper mines and concentrators). However, preservation of the biodiversity and ecology of the desert is key. To develop this region, infrastructure (port, rail, road, air and water) and energy (gas, hydropower, wind and solar), potable water (desalination of seawater) and ecological conservation are needed. In applying the DSP approach, we can identify a phased approach, which is laying out a path with clearly laid steps, for developing a diverse and sustainable industry for the region.

Step 1: Industrial Diversification Design

For each identified source of raw material, the quality, quantity, location and ease of extraction are assessed. The process routes required to convert the resources into products that can be shipped to users are then defined, including a sustainable supply, energy and water and the added value of producing in Peru and shipping versus just shipping the raw material is examined to enable rational decisions on which processes to install. If more of the process tasks are kept in Peru, the wealth of the country and the people would increase; the first step toward developing downstream manufacturing. Further, the synergies between different processes and industries must be considered. By-products and waste from one industry can be repurposed for another industry, such that limited waste is generated and the desert ecology is preserved. Integration of industries into a Circular Economy is key to collaborative development and environmental preservation.

For example, instead of shipping copper concentrate, metallic semi-finished copper in the form of cast slabs or billets could be manufactured. Further value could be added to the product by processing it into rods, wire or sheet metal. The latter processes could be considered as medium or long term activities. Similar opportunities exist for phosphates and potash (e.g., downstream production). Fertilisers could be manufactured rather than shipping the raw materials, which would create an opportunity to build a large agricultural and food industry. Fertiliser manufacturing, coupled with the suitable climate of the area, desalination or reprocessing of industrial water, could create an opportunity to build a large agricultural and food industry in the region. Further, the diatomite and calcareous gypsum resources, working in conjunction with copper refineries and reusing their slag by-product, can support the local cement industry.

Step 2: Sustainable Expansion

The development of a Circular Industry Cluster around a copper mining and refining operation, fertiliser plants, cement plants and associated agricultural and food industries has knock-on effects. Education programs would be needed to upgrade the skills of the local workforce. In assessing this potential benefit, environmental and social sustainability impacts must be assessed. This also applies to other industries being considered.

The benefits of the Circular Industry Cluster can be used to attract major investments in the area. The planned cluster should also outline how to promote the benefits and sustainability to encourage the long term formation of large, medium and small company associations. Suggestions should be developed for other industries to develop and diversify the industry base to provide a stable economy. For example, the fishing industry can first start with fishing from local waters, followed by expansion of sustainable fish farming and then extending to value added fish canning. The sustainability of each potential industry is to be assessed within the context of resource availability, environmental and socio-economic impacts and social or political risks and opportunities.

Step 3: Education, Training and Socio-Economic Development

By defining the types of skills required to work in the industrial operations being considered, educational and training targets should be identified. This is aimed at providing a basis for strengthening the educational system for the area and ultimately leading to phases of expansion such as:

- **First Phase:** General education and development for unskilled workforces.
- **Second Phase:** Skills development for future industries.
- **Third Phase:** Advanced industry/university programs to study new developments. For example, in the fertiliser and fishing industries (fish banks, investigating marine species, farming, etc.). This may include partnering with global leaders in some of these niche areas. As part of this assessment, the socio-economic profile of the region should be reviewed to determine the number of people needed to sustain the expansion and population availability and mobility. Ideas for engaging the local population to identify potential risks and opportunities to be considered. This includes an assessment of local assets, which will identify opportunities for building on existing socio-economic systems. An inclusive transition and development plan can then be developed that looks at transitioning the region into a sustainable diverse economy.

This stepwise approach would ensure an inclusive transition of the local workforce to support the region's sustainable diverse economy. Areas that are likely to have a major impact on the success of the region should be analysed, for example:

- Market and prices
- Raw materials, energy and major consumables
- Process technologies and operations
- Country, region, infrastructure and services
- Environment
- Social and socio-economic factors
- Legal framework.

A possible outcome is a sustainable Circular Economy developed over 10–15 years, comprised of:

- Establishment of the Peruvian markets for fertilisers and copper products
- Meeting environmental targets and social needs in a sustainable manner

- Opportunities for future development and value added products in fertilisers, agriculture products, food industry, copper products and cement products
- Sustainable development
- Attraction of participation and investment from major corporations in the local economy
- Formation of synergistic associations of large, medium to small companies
- Education, training and human development to support the future industries
- Creation of a comprehensive infrastructure including transportation, energy generation and supply and water management including wastewater from industry.

References

Andersen, R. F. (2017). An overview of the progress and challenges of peatland restoration in Western Europe. *Restoration Ecology*, 25(2):271–282.

Atlas of Economic Complexity. (2023). What did Peru export in 2019? Retrieved from https://atlas.cid.harvard.edu/explore?country=173&product=undefined&year=2019&productClass=HS&target=Product&partner=undefined&startYear=undefined

Bebbington, A., & Williams, M. (2008). Water and mining conflicts in Peru. *Mountain Research and Development*, 28(3):190–195.

Bourgault, M.-A., Larocque, M., & Garneau, M. (2019). How do hydrogeological setting and meteorological conditions influence water table depth and fluctuations in ombrotrophic peatlands? *Journal of Hydrology X*, 4:100032.

Bruch, A., Patchigolla, K., Asfand, F., Turner, P., Monte, L. M., & Douard, S. (2020). Industrial scale cTES cold thermal energy storage: Demonstrator in La Africana CSP power plant and evaluation of benefits – SOLWARIS project. *SOLARPACES 2019: International Conference on Concentrating Solar Power and Chemical Energy Systems* (October 1–4, 2019). Daegu.

Canada Energy Regulator. (2023, August 24). *Provincial and territorial energy profiles – Quebec*. Retrieved from Canada Energy Regulator: https://www.cer-rec.gc.ca/en/data-analysis/energy-markets/provincial-territorial-energy-profiles/provincial-territorial-energy-profiles-quebec.html

Economist Impact. (2022, February 10). *Data point: a wet, wild, wonderful world*. Retrieved from Economist Impact: https://impact.economist.com/sustainability/ecosystems-resources/data-point-are-wetlands-are-the-original-carbon-capture-and-storage-systems

Encyclopedia Staff. (2023, June 21). *Colorado Gold Rush*. Retrieved from Colorado Encyclopedia: https://coloradoencyclopedia.org/article/colorado-gold-rush

ENEA. (2017). *MATS Project*. Retrieved from Italian National Agency for New Technologies, Energy and Sustainable Economic Development: https://www2.enea.it/en/enea-atlas-for-development-cooperation/project/mats-2013-multipurpose-applications-by-thermodynamic-solar

Girard, M., Lavoie, C., & Theriault, M. (2002). The regeneration of a highly disturbed ecosystem: A mined peatland in Southern Québec. *Ecosystems*, 5:274–288.

Goodbrand, A., Westbrook, C. J., & van der Kamp, G. (2019). Hydrological functions of a peatland in a boreal plains catchment. *Hydrological Processes*, 33(4)562–574.

Grasby, S. E., Allen, D. M., Bell, S., Chen, Z., Ferguson, G., Jessep, A., ... Therrien, R. (2012). *Geothermal Energy Resource Potential of Canada*. Geological Survey of Canada.

Harris, L. I., Richardson, K., Bona, K. A., Davidson, S. J., Finkelstein, S. A., Garneau, M., ... Ray, J. C. (2021). The essential carbon service provided by northern peatlands. *Frontiers in Ecology and the Environment*, 20(4):222–230.

Hiam-Galvez, D., Gleming, H., & Osores, O. (2012). Water supply in Peru & Chile, the challenges and solutions. *3rd International Congress on Water Management in the Mining Industry* (June 6–8, 2012). Santiago Chile.

Metro Denver Economic Development Corporation. (2018). Metropolitan Denver Region Industry Clusters.

Metro Denver Economic Development Corporation. (2022). *Initiatives*. Retrieved from https://www.metrodenver.org/about/initiatives

Miller, B., & Sweigart, E. (2019, October 15). *How countries manage water: Peru*. Retrieved from Americas Quarterly: https://www.americasquarterly.org/article/how-countries-manage-water-peru/

Ontario. (2022, March 25). *Ontario's Ring of Fire*. Retrieved from https://www.mndm.gov.on.ca/en/ring-fire

Oxford Business Group. (2022). *Mining, from the report: Peru 2017*. Retrieved from https://oxfordbusinessgroup.com/peru-2017/mining

Poirier, K., Al Mhanna, N., & Patchigolla, K. (2022, October 8). Techno-economic analysis of brine treatment by multi-crystallization separation process for zero liquid discharge. *Separations*, 9(10):295.

Ramsar Convention on Wetlands. (n.d.). *Global wetland outlook*. Retrieved from https://www.global-wetland-outlook.ramsar.org/

Richter, A. (2020, February 14). *Geothermal energy could contribute up to 50% to Peru's electricity supply*. Retrieved from Think GeoEnergy: https://www.thinkgeoenergy.com/geothermal-energy-could-contribute-up-to-50-to-perus-electricity-supply/

Rochefort, L., Quinty, F., Campeau, S., Johnson, K., & Malterer, T. (2003). North American approach to the restoration of Sphagnum dominated peatlands. *Wetlands Ecology and Management*, 11:3–20.

Sansom, C., Patchigolla, K., Jonnalagadda, K., & King, P. (2020). Design of a novel CSP/MED desalination system. *26th SolarPACES Conference 2020: Solar Power & Chemical Energy Systems*. Virtual Event.

Shire of Laverton. (2022, September 3). *Shire of Laverton*. Retrieved from https://www.laverton.wa.gov.au/

Shire of Leonora . (2022, September 3). *Shire of Leonora*. Retrieved from https://www.leonora.wa.gov.au/visitors/about-leonora/about-us.aspx

Smolders, A. J., Tomassen, H. B., Lamers, L. P., Lomans, B. P., & Roelofs, J. G. (2002). Peat bog restoration by floating raft formation: the effects of groundwater and peat quality. *Journal of Applied Ecology*, 39(3):391–401.

Solar Water Plc. (2023). *Sustainably solving the water crisis*. Retrieved from Solar Water: https://www.solarwaterplc.com/

Statista. (2022a). *Copper mine production in Peru from 2009 to 2020*. Retrieved from https://www.statista.com/statistics/789003/copper-production-peru/

Statista. (2022b). *Employment in the agricultural sector in Peru from 2010 to 2019, as share of total employment*. Retrieved from https://www.statista.com/statistics/1080919/peru-share-employment-agriculture/

Tanneberger, F. A. (2021). The power of nature-based solutions: How peatlands can help us to achieve key EU sustainability objectives. *Advanced Sustainable Systems*, 5(1):2000146.

Tarnocai, C., Kettles, I. M., & Lacelle, B. (2011). *Geological survey of Canada*. Retrieved from Peatlands of Canada: https://doi.org/10.4095/288786

The Interior News. (2021). The Interior News. Retrieved from Northwest B.C. mining and exploration summary for 2020: https://www.interior-news.com/business/northwest-b-c-mining-and-exploration-summary-for-2020/

Trading Economics. (2022). *Employment Rate in Peru increased to 91.70 percent in April from 90.60 percent in March of 2022*. Retrieved from https://tradingeconomics.com/peru/employment-rate

United Nations Environment Programme. (2023, February 27). *Critical ecosystems: Congo Basin peatlands*. Retrieved from United Nations Environment Programme: https://www.unep.org/news-and-stories/story/critical-ecosystems-congo-basin-peatlands

Vaisala. (2017). *Vaisala 3TIER Services Global Solar Dataset (Ref. B211641EN-B)*. Retrieved from https://www.vaisala.com/sites/default/files/documents/3TIER%20Solar%20Dataset%20Methodology%20and%20Validation.pdf

Western Australian Museum. (2022, September 3). *Gnamma Holes*. Retrieved from https://museum.wa.gov.au/explore/wa-goldfields/water-arid-land/gnamma-holes

World Wildlife Fund. (2023). *Scotland's amazing peatlands*. Retrieved from World Wildlife Fund: https://www.wwf.org.uk/scotlands-amazing-peatland

9

Education, the Lighting of a Fire

Anthea Kong[1], Jake Wyman[2], and Doris Hiam-Galvez[3]

[1] Hatch Ltd, Industrial Clean Technologies, Mississauga, Ontario, Canada
[2] Hatch Ltd, Renewable Power and Electrical Distribution, Mississauga, Ontario, Canada
[3] Hatch Ltd, Designing Sustainable Prosperity, Vancouver, British Columbia, Canada

> *"Education is not the filling of a pail, but the lighting of a fire."*
>
> *– W.B. Yeats*

> *"The foundation of every state is the education of its youth."*
>
> *– Diogenes*

Diogenes' remarks also apply to DSP since the foundation of building up DSP is education. Building a sustainable society starts from the ground up, metaphorically. When it comes to the development of a society, where better to start than in the education system and schooling?

Early Schooling

Bees and school children (Figure 9.1) usually are not considered a good mix. But what if this combination gave way to possibilities that were never considered before? This is what neuroscientist Beau Lotto did – put bees and school children together to create an environment that nurtured some of the youngest published scientists ever, at just 8–10 years of age, proving that children are capable of high levels of thinking.

Picture 'school'. Is the image an array of students facing the front of a room where a teacher is lecturing? Are there tests and assignments that are rewarded if done well and punished if mistakes are made? This is the image most often depicted of school and learning, but it does not give students the tools needed to function and live in the real world. After all, isn't that what school is for: to prepare the youngest members of the population for 'real life'? To learn means to understand, to think and to cooperate with others.

Designing Sustainable Prosperity: Natural Resource Management for Resilient Regions,
First Edition. Edited by Doris Hiam-Galvez.
© 2024 John Wiley & Sons, Inc. Published 2024 by John Wiley & Sons, Inc.

Figure 9.1 Bees and school children.

Fundamentals

Most problems that life presents cannot be solved just by sitting in a lecture hall. However, foundational knowledge is required before building skills to solve problems. These include mathematics, science, arts, health and interpersonal skills.

Money plays a vital role in the functioning of the world, yet how often are finances widely taught in school? We cannot expect children to grow up instinctually knowing what is a good investment and how to save money. Hand in hand with finance is statistics, which is necessary to understand the workings of the world and the variability that inherently exists. A basic understanding of science would also help in understanding the world better. If framed as a discovery of the world and exploration of life, science would be more engaging and attainable. A game-like approach to science would be even more appealing to children, as described by Beau (Lotto, 2016). More on this approach is explored in subsequent sections.

The arts allow for an opportunity to explore the world through a different lens and to learn from others, as well as about ourselves. Studying art fosters a creative mind required to innovate solutions when combined with problem solving skills and scientific knowledge. Health and fitness is a subject that is often structured to be neglected in the education system. As one advances in the education system, the importance of health and fitness wanes. By the time one exits the education system, it is often developed into a burden that adults must take on. Stressing the importance of one's health can help alleviate some of these resulting burdens.

Humans are social beings – we live and thrive in communities and in the company of one another. As such, interpersonal skills are very important. Communication is a two-way interaction. We must present our thoughts in the 'language' of the receiver to avoid a breakdown in the bridge between the presenter and the receiver. Listening and reading are equally as important. Often, our society does not put enough emphasis on the art of listening. Usually, we listen to others to form an answer or to validate our own assumptions and views, instead of listening to what the other is trying to share without the addition of our own opinions. By truly listening to others, we are given a chance to expand our perspectives and view of the world and to understand others better, which leads to more inclusive thinking, empathy and compassion. Only when both the receiving and expressing are executed well can there be collaboration between different individuals – leading to a deeper understanding of one another. With collaboration and genuine listening, we can better achieve UN SDG 17, where partnerships are necessary for the achievement of the SDGs.

Encouraging Curiosity

> *"Curiosity is the wick in the candle of learning."*
>
> *– W.A. Ward*

How often have you considered how something works? In Nadya Mason's TED talk (Mason, 2019), she explored the idea of learning how the objects of our everyday lives work. The example she gave was of smartphones. Objects upon which – for many of us – our lives depend. Yet, most people do not have the slightest idea of how they work. This extends to other appliances, toys and even our manufactured food. Consider cotton candy. Most know what cotton candy is and have seen it – if not eaten it – many times in their lives before. Yet, how is it made? What are the basic physical and chemical principles at play to make this seemingly magical, fun treat? This is a basic example but having this kind of curiosity will widen perspectives of everything around us, including how we relate and affect them. How do hummingbirds fly backwards, sideways, even upside down? Why do airplanes often have a white trail behind them when they fly? How do trees take up water hundreds of metres into the sky?

By being curious about the world around us, we can better connect with it. By designing education to foster growing curiosity, the learner develops a lens that allows them to question and make sense of the world and hence connect with it. Only through connection can a relationship be developed. And only through relationship building can care and stewardship be stimulated – towards each other and the world.

There are various ways to explore and satisfy this curiosity. One can look up information online and watch videos. Another way is to experiment. One is quick and easy, the other requires more time, energy and resources to implement. However, the latter needs to be encouraged more often to truly develop perpetuating curiosity and to generate critical thinking skills, as opposed to retaining information skills. This leads to the requirement for more interactive and experimental styles of teaching and learning.

An example of this is the Blackawton bees paper (Blackawton, et al., 2010) mentioned at the opening of this chapter and shown in Figure 9.2. Led by Beau and Dave Strudwick, the paper was executed and written – for the most part – by school children 8–10 years of age. This was a breakthrough example wherein a group of school children proved to be capable of higher-than-expected levels of thought processing, by framing science in the manner of a game – to obtain a goal given a set of rules. This kind of activity gives permission for children to explore for themselves, connect with the ecology that interacts with them and develop the foundation for them to love science and find it fun since a game is playing – something children enjoy.

> *"The process of playing with rules enables one to reveal previously unseen patterns of relationships that extend our collective understanding of nature."*
> *– Beau Lotto (Blackawton, et al., 2010).*

Figure 9.2 School children working with Beau Lotto on the study of bees. *Source:* Dr. R. Beau Lotto.

Another way to encourage curiosity is to advocate for children to do hands-on work. In many parts of the world, societies have partitioned themselves from doing hands-on work. It has a negative connotation to it – it is too messy, it takes too long, it is inconvenient; it results in lots of error, destruction and uncertainty. Since education is built on a model of getting things right, the idea of making a mistake is negative, leading to students learning early that if they do not know how to do something, it is better to just not try. This mistake aversion also leads to a lack of innovative and creative development since these processes require mistake-making. When they grow up, these students will not know what are proper risks to take, what kind of mistakes lead to detrimental consequences and which mistakes are okay to make. Hands-on work develops a child's judgement skills, not only in risk assessment but also in the spatial sense. By physically seeing and working with objects, spatial relativity and judgement can be instilled at a young age, accumulating more experience in predicting and estimating physical possibilities.

When we extend the encouragement and valuing of hands-on work to later stages of education, we can build a more diverse workforce – one in which labour and trade are respected. The idea of learning in the classroom being combined with learning in the workplace is being implemented sporadically in various places – with the popularity of internships and co-op education in post-secondary education programmes. However, it is important that we also apply this in earlier stages of education to promote diverse educational development. One example is Germany's education system, where the development of practical skills is widely supported as part of an option in secondary school. Germany has long offered options in apprenticeship, vocational education and dual education programmes (Trines, 2021). In these vocational programmes, which constitute an estimated amount of about half of student enrolment, manual work and trade skills are developed with the involvement of local businesses – supporting the needs of the local economy while also equipping the young population with practical skills on top of theoretical knowledge. This kind of system also helps with dispelling a hierarchical mindset that manual work and trade skills are second class. Instead, this workforce is equally respected as a vital part of diverse industries and the economy. In some ways, this system promotes SDG 8 since it enables various kinds of work that would suit different people. More on how education can be modified to support future industries in the DSP framework will be discussed in subsequent sections of this chapter.

Connecting with Nature

As a modern society, we have been developing in the direction of increasingly separating ourselves from nature. This is taught early on, in school, where we do not encourage the exploration of the outdoors and nature. Growing up, we often learn that we must conquer nature – both implicitly and explicitly, that we must fight it, that it is exploitable and not something that we should cherish. As a result, in the times when we do want to cherish it, we are not equipped with the right tools.

By connecting with nature, we are better able to see the world as a whole system, with ourselves as part of it. When we are a part of something, compassion and curiosity are elevated. Currently, as humans, we are under the impression that we are superior beings and that we have a right to own and exploit the environment around us. Then we try to fight nature to solve our problems. Yet, nature is a balanced and circular ecosystem, one in which there is no waste and no discontinuity. Humans are a part of this ecosystem, whether we like it or not. We must take responsibility to figure out how we fit in this ecological community. Fitting into the community includes knowing how to consume and produce responsibly, as outlined in SDG 12. When we see ourselves as a part of the community, perhaps we will learn what it means to be stewards of the environment.

We also must learn from nature. And why not? The beings that surround us have vast amounts of experience. Take spiders, for example, as shown in Figure 9.3. Capable of incredible engineering by spinning intricate webs with material that is strong and sticky – both properties vital to their survival. There are also trees, ants, hummingbirds, frogs, dandelions, lilies, corpse flowers, moths, salmon, etc. The list can go on because there is much inspiration that can be borrowed from nature if only we knew how and were encouraged to learn from them.

Light guides us on how to see, not dictates what to see.

Figure 9.3 Imagine the intricate process of creating this scene. *Source:* ArtStage/Adobe Stock Photos.

Many education systems approach learning in a one-way teaching method that is based on information memorisation, exam-passing and getting things right. To equip the next generation to function in society effectively, there needs to be a change in how they develop these tools. Approaching the learning experience with fun, curiosity, mistake encouragement and hands-on discovery will enable connection with the world and nature, allowing for better understanding and judgement (Figure 9.4). A few examples of teaching using these innovative ways were mentioned, but they need to be implemented more broadly in education systems and in various regions. With SDG 4 in mind, regions developing new systems will present the perfect opportunity to implement these methods. With this new approach to education in mind, it is necessary for educators to learn and accept that the learner must actively participate in the learning process with the educator as a guide. It may be difficult for educators to not be in control since that is how education has been up until now. This is where we all must learn to change our perspective of what education looks like. The process of relearning and retraining is not simple but is required for the development of an improved, sustainable future.

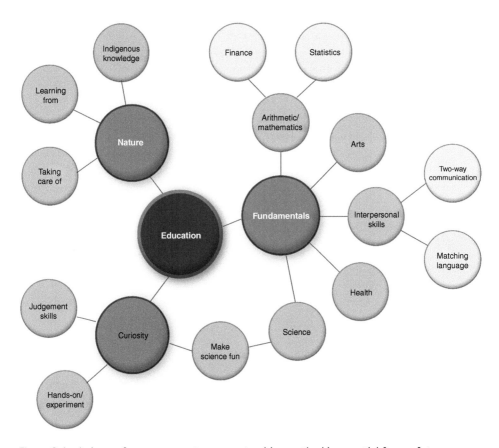

Figure 9.4 A change from our current one-way teaching method is essential for our future.

Retraining and Upskilling

Trying to escape a house fire using a flame is futile. So, why do we try to improve our world by using the same actions that were used to damage the world in the first place? Why have we triggered climate change by burning oil and natural gas but continue to train current generations the skills to continue with these actions? Why do we complain about wasting precious natural resources but continue to buy showerheads that waste twice as much water?

A more sustainable future is dependent upon the education we bestow upon the current generations today. We should not teach the skills that were necessary in the past, but we must rather focus on the skills that are essential for the industries of the future. With the DSP framework, we can effectively determine the necessary skills to support a more sustainable future and reduce skill mismatches to create long term economic and ecological prosperity.

Just as plants require energy from the sun, a supporting stem, nutrient-rich soil and water to survive, a more sustainable future must have industries such as ample renewable energy generation, strong supporting services, sustainable food production and freshwater preservation (Figure 9.5). Consequently, the workforce's current skillset must be adjusted to match this view of the future. As discussed earlier, childhood education must undergo changes in both the content and how it is delivered. However, the workforce's skillset can also be adjusted through later life learning with retraining programmes. Retraining programmes play an essential role in creating sustainable prosperity by providing the necessary skills to individuals to succeed in long term careers rather than short term gig work.

Figure 9.5 Growing a sustainable future by nurturing renewable energy generation, strong supporting services, sustainable food production and freshwater preservation.

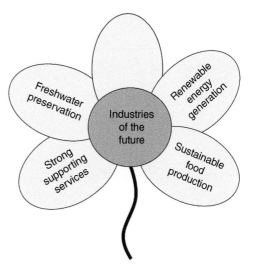

An example of using retraining programmes to achieve the objectives of DSP is the retraining of Peruvian workers in the copper industry. Peruvian copper workers have developed advanced thermodynamic and heat transfer skills in copper smelting plants. Instead of leaving these workers to pursue alternative jobs upon mine closures, retraining programmes can be implemented to build upon their existing experiences with processing heated fluids and assist workers in transitioning to a budding renewable energy industry. Peru's high solar irradiance is primed for the use of concentrating solar power (CSP) technology, where sun rays are redirected using mirrors to heat water into steam, with the steam being used to turn turbines and generate energy (Ruiz, 2019). The Peruvian copper smelters' experiences with high temperature liquid copper will enable them to iterate and improve upon the efficiencies of the CSP technology and reduce Peru's dependence upon fossil fuels.

Similarly, the thermodynamic skills practised in the copper industry can be used to increase the viability of geothermal energy generation by using a supercritical fluid to increase the efficiency of the heat transfer to more closely approximate the ideal Carnot cycle. This will increase the feasibility of implementing geothermal energy generation since the capital investment will yield greater energy generation and economic returns. Through the implementation of specific retraining programmes, Peruvian mining workers' existing skills can be supplemented to enable a transition to a similar role that will both increase the workers' long term job security and ensure a more sustainable future.

> *"Anyone who stops learning is old, whether at 20 or 80. Anyone who keeps learning is young."*
>
> *– Henry Ford*

In life, there is no graduation date where you have suddenly learned all the knowledge that you will need for your lifetime. Regardless of your age, you are always learning new ideas, meeting different people and obtaining new experiences. The understanding that education is not periodic but rather continual must be applied to the workforce as well with workers needing to continually learn to adapt to the constantly changing world around them.

This is where upskilling plays a crucial role in assisting individuals to improve their skills and adopt the most recent and sustainable technologies. Farmers should be completing upskilling programmes to learn how to use the most recent technologies to create more sustainable agricultural practices. The Netherlands has created agricultural technology that monitors soil moisture to avoid excess water usage from the overwatering of crops (Sensoterra, 2024). Farmers may not naturally know how to use this technology, but with the use of upskilling programmes, they can learn how to implement this innovative technology and achieve a more widespread reduction in water usage.

> *"If we teach today's students as we taught yesterday's, we rob them of tomorrow."*
>
> *– John Dewey*

Retraining and upskilling programmes are instrumental in creating a more prosperous future, yet they must be operated properly to succeed beyond simply symbolic actions. First, these programmes can only be effective if they address the skillset demand in the nearby regions. Implementing nuclear energy retraining programmes is futile if there is no local nuclear energy industry for the participants to enter after graduation. Second, workers must be prepared for the industries of the future instead of those of the past. Retraining workers to participate in the oil and gas industries both reduces their long term economic stability and slows society's transition to renewable energy generation.

The previous examples showcase the ways in which we must adapt our education model to support sustainable prosperity, with the specific methods and results differing based upon the region being studied. Nevertheless, the consistent themes across all regions are the need for forward-looking education curricula as well as the value of lifelong learning through retraining and upskilling programmes (Figure 9.6). The various points highlighted will be required for all regions to build towards a sustainable, future society.

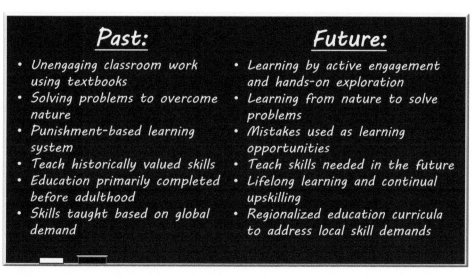

Figure 9.6 Adapting our past education model to support future sustainable prosperity and development of a future society.

References

Blackawton, P.S., Airzee, S., Allen, A., Baker, S., Berrow, A., ... Beau Lotto, R. (2010). *Blackawton bees*. Retrieved from Biology letter, The Royal Society: http://rsbl.royalsocietypublishing.org/content/suppl/2010/12/21/rsbl.2010.1056.DC1.html

Trines, S. (2021, January 28). *Education in Germany*. Retrieved from World Education News and Reviews: https://wenr.wes.org/2021/01/education-in-germany-2

Lotto, B. (2016). *Deviate: The Science of Seeing Differently*. Hachette Books.

Mason, N. (2019, October). *Nadya Mason: How to spark your curiosity scientifically*. Retrieved from TED Talks: https://www.ted.com/talks/nadya_mason_how_to_spark_your_curiosity_scientifically?referrer=playlist-the_pursuit_of_curiosity_and_understanding

Ruiz, F. P. (2019). *Concentrating Solar Power Potential of Peru*. Aalborg: Aalborg University.

Sensoterra. (2024). *Precision Agriculture & Irrigation Control*. Retrieved from Sensoterra: https://www.sensoterra.com/use-cases/agriculture-horticulture/

10

Future Society

Matthew Tutty[1] and Doris Hiam-Galvez[2]

[1] *Hatch Ltd, Climate Change, Mississauga, Ontario, Canada*
[2] *Hatch Ltd, Designing Sustainable Prosperity, Vancouver, British Columbia, Canada*

The DSP model provides a method to achieve a modern society that is sustainable.

This chapter discusses what a future society should look like, illustrated by Figure 10.1, with examples including some from the past; for example, Leonardo da Vinci's vision of a society that was inspired by nature's design and oriented to well-being.

Figure 10.1 Leonardo da Vinci's ideal city. *Source:* ShuaiGuo / Pixabay.

Societies are a critical aspect of the human experience by forming our childhoods, connecting us with work and relationships and providing us with vital resources. Therefore, the prosperity of humans and our planet is tied to the prosperity of our societies. Unfortunately, there is usually a gap between the ideal society and reality. Overconsumption, destruction of ecosystems, rising inequality and other vulnerabilities in our communities are seen every day. A utopian, future-ready and sustainable society can co-create value economically, socially and environmentally through proper planning. Engineering plays a role in designing circularity and, together with science, arts and culture, fosters innovation. In contrast with current practice, future societies must be linked with nature in ways that generate value. Values-based decision-making will create sustainable consumption, increase income and social equality, promote ethical technology development and more responsible corporations and generate needed innovations. With this focus, we can develop future-ready and sustainable societies.

News stories are most commonly about a devasting natural disaster, an act of extremism, or a lived experience of poverty or crime, with few stories that highlight actions to counteract these disparities. In our cities, we are generating waste that cannot be recycled properly or locally treated because the rate of generation is too high or the waste is not recyclable. Hunger and poverty still exist despite record levels of wealth generation amongst the richest members of the global economy. Such imbalances have spilled into the natural world as well. Urban sprawl and unsustainable farming practices are destroying our planet and jeopardising or have already destroyed the habitats of many plants and animals. How societies operate contributes to these issues. For example, in many cities, there is a lack of comprehensive public transport, little green space and inaccessible local amenities. These problems interact and have compounding negative consequences. The interconnectivity of these issues exacerbates their impacts but also provides an opportunity to improve many aspects of society through addressing the root problems. Dreaming up a more sustainable society is not a new concept. Leonardo da Vinci worked to develop an urban utopia through drawings and models. This perfect city included locks to make the waterways efficient and avoid floods, an emphasis on hygiene and optimised transportation networks. Efficient transportation of goods was designed into the city (Melis, 2019). Architectural beauty and comfortable spaces were planned. His city had nature incorporated into the design with waterways throughout for commerce and he emphasised building vertically, not horizontally, to save space. The deliberate intersection of engineering, architecture, urban planning and nature allowed for innovative solutions to address the problems experienced in cities at the time. Further, prosperity and the well-being of the city's population were central to da Vinci's master plan. Da Vinci's utopia was never constructed but foreshadowed and inspired cities to come. Utopias described in history, literature and film are underpinned by harmony between the many aspects of society. Leonardo da Vinci's ideal city showed us the importance of using engineering to design connected spaces where the various facets of a society work together to promote collective well-being.

Is a Utopia Possible?

"Learn how to see. Realise that everything connects to everything else."
– Leonardo da Vinci

What aspects of society today need to change to lead to a better future? How do we close the gap from our current reality to a future-ready world? Societies globally and regionally can be quite diverse in their construction. To design a connected and values-based society, as da Vinci had imagined, the multiple physical and non-physical aspects must be understood.

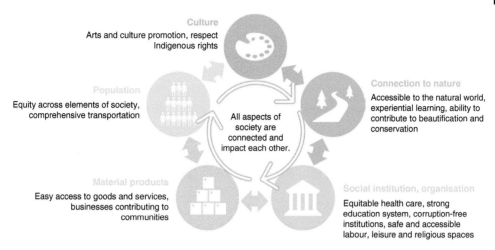

Figure 10.2 The multiple elements of society interrelate and can build upon each other to develop solutions that benefit all key parties.

Multiple elements of a society can be categorised into culture, population, material resources, social institutions, social organisation and connection to nature, as shown in Figure 10.2.

These segments can either add to or take away from the prosperity of the community. Therefore, each segment can be analysed for improvement and a solution can be engineered across multiple elements to create an overall benefit. A caution with da Vinci's work is that it is often not possible to design a new community where there was none before and expect buy-in. Rather, designing a future-ready society usually requires improving an existing one.

Highlighting the Past

The Chavín people resided in the highlands of Peru from 900 BCE to 200 BCE (Burger, 1990). They were innovative in adapting their structures to the complex highland environment of the Andes mountains. For example, the Chavín de Huántar temple had a functioning drainage system with canals below the temple that removed rainwater away from the temple. The Chavín people had impressive developments in acoustic engineering. A network of resonance rooms was connected by sound transmission tubes (Kolar, 2010). During rainy seasons, water would rush through the canals of temples and create a noise like the sacred jaguar (Lumbreras, 1976). Also, the Chavín people are known for improvements to agricultural practices and the development of metallurgy and textiles (Lothrop, 1951). The Inca civilisation, following the Chavín culture, existed between 1438 and 1533 in Western South America (New World Encyclopedia, n.d.). The Inca culture faced a similarly challenging and harsh environment as the Chavín people, making agriculture and transportation difficult. The Incas are famous for their architecture and ability to adapt to the natural world where they lived. Most prominently, their terraced societies like Machu Picchu were constructed to develop communities and agriculture in the extremely rainy climate and hilly terrain. The soil of these Andes Mountains has been cultivated for at least 1500 years and is still productive, a testament to the Inca's understanding of soil management and sustainable agricultural practices. Recently, there has been a restoration of abandoned Andean terraces, leading to economic opportunities for local communities (Brevik, 2018).

Ancient East Asian cultures are also recognised as having excellent sustainability practices through purposely selected crops for the climate, engineered water management systems, use of mulches for water conservation and the use of various materials as fertiliser. Additionally, these practices invoke notions of a circular economy where no organic waste is left unused, instead consumed, worn or used as a fuel or fertiliser. These practices, along with a very favourable climate for agriculture, have allowed for consistent cultivation for 4000 years (Brevik, 2018).

An example of the beginning of a sustainable society can be observed in Nabatean Petra, a city built in the desert in 300 BC with limited water resources and complex geology. They lived there for seven centuries and at one time, it was home to 30K people. Nabateans, to make this possible, used a high degree of hydrology mastery for the exploitation of all possible water resources, including unique management techniques, to balance reservoir storage capacity with continuous flow pipeline systems to maintain a constant water system throughout the year. The design of the water network system promoted stable flows and used sequential particle settling basins to ensure potability. Open channels flow within piping to stay below critical flow rates were used to avoid leakage during times of high flow rates in the spring (Ortloff, 2005). The combined result of this technology was a society that survived for a long time in harsh conditions.

These ancient cultures, as with many Indigenous cultures globally, had a deep respect for their natural surroundings and worked with local resources to their advantage to develop innovative solutions for the individual community's needs. These societies had an intimate understanding of the area's resources which were used to benefit society and be a driver of innovation. We must work with the natural world to simultaneously provide prosperity for current and future generations. What unrealised benefit could we experience if we connected with nature in the design of our societies? Also, these ancient societies utilised circularity which must now be central to stopping overconsumption and waste. In a sustainable society, everyone has what they need without the generation of waste, thereby promoting circularity and equality. It is noted that history can idealise civilisations and may not capture the realities for all those in a region. These ancient civilisations lacked in many areas and had small and relatively isolated populations, not comparable to modern society. Therefore, inspiration can be drawn from da Vinci's urban utopia and the ancient civilisations previously mentioned, but in the end, our future-ready society must be completely novel to overcome the present disparities experienced today.

A highly interconnected society that utilises values-based leadership and innovative engineering and is connected with nature sounds amazing but is it achievable? No society is perfect, but a collaborative effort with all parties working with local communities will develop buy-in to design a more sustainable community for all. Specifically, the ancestral values of the local people should always be considered and respected with attention to balancing new ideas and observing traditions and local values.

What may be considered a sustainable society for one region may not hold for another. Common themes of a sustainable society would be sensible wealth that is spread equitably and where the highest earners are not acquiring wealth at the expense of the lowest earners. The health and well-being of humans and nature are prioritised through the multiple aspects that impact health. In a sustainable society, there must be a deep connection with nature, as was seen in ancient civilisations. There must be sustainable means to produce food and necessary natural resources. Connection to nature must also occur at the personal level. There should be places within communities to enjoy and spend time with nature. Transportation

Where we are now	Where we want to go
Silo-ed aspects of society result in inefficient use of resources	Highly interrelated society
	Connection to nature
Unsustainable consumption	Curiosity and innovation spark novel solutions
Divisive politics and cultures	Accessibility and equality are central in decision-making
Not connected to each other and to nature	Environmentally sustainable society

Figure 10.3 Where we want to go.

should be accessible and retail spaces placed in the communities they serve. Furthermore, businesses, large and small, contribute to the local economy through local sourcing, outreach and engagement. Citizens should have meaningful, safe and fulfilling work. Overall, a sustainable society is connected, sustainable and prosperous in every dimension. DSP enables the creation of future sustainable societies where there is natural resource extraction activity. These sustainable communities can serve as templates for the region and spread prosperity regionally. In this way, prosperity can be realised without sacrificing sustainability. What aspects of society does your project contribute to and how is your project contributing to a future-ready and sustainable society? What actions are you taking in the development and execution of your project to leave a lasting positive impact on the society (see Figure 10.3)?

Inspiration for a sustainable society can be found in nature. Natural ecosystems have been developing for thousands of years and, when left alone, are highly functioning communities. Co-habitation of multiple species of animals and plants benefits from each other up and down the food chain. Animals use plants for food, shelter and nesting and in turn provide fertiliser, pollination and species management. The values promoted in DSP are evidenced in nature and can be used as an input for the design.

References

Brevik, E. C. (2018). Soils, climate, and ancient civilizations. *Developments in Soil Science*, 35:1–28.

Burger, R. L. (1990). Maize and the origin of highland Chavín civilization: An isotopic perspective. *American Anthropologist*, 92(1):85–95.

Kolar, M. A. (2010). A modular computational acoustic model of ancient Chavín de Huántar, Perú. *The Journal of the Acoustical Society of America*, 128(4):2329.

Lothrop, S. K. (1951). Gold artifacts of Chavin style. *American Antiquity*, 16(3):226–240.

Lumbreras, L. G. (1976). *Acerca de la función del sistema hidráulico de Chavín*. Museo Nacional de Antropología y Arqueología.

Melis, A. (2019). *Leonardo da Vinci designed an ideal city that was centuries ahead of its time.* Retrieved from The Conversation: Leonardo da Vinci designed an ideal city that was centuries ahead of its time: theconversation.com

New World Encyclopedia. (2018). *Inca Civilization.* Retrieved from New World Encyclopedia: https://www.newworldencyclopedia.org/entry/Inca_Civilization#:~:text=%20Inca%20Civilization%20%201%20Emergence%20and%20Expansion.,Inca.%20When%20a%20new%20ruler%20was...%20More%20

Ortloff, C. (2005). The water supply and distribution system of the Nabataean City of Petra (Jordan), 300 BC – AD 300. *Cambridge Archaeological Journal*, 15(1):93–109.

Embracing Change: A Call to Action

As we conclude this transformative journey, a clear call to action emerges. These final pages urge us to join hands in impactful efforts that will shape a sustainable future.

Throughout this book, you have learned of a structured framework and practical steps to be an active participant in this transformation. With knowledge and insights, you are well prepared to thrive in the ongoing journey of learning. Community empowerment and a new way of thinking are key to our shared path forward.

Designing Sustainable Prosperity (DSP) is the key to unlocking untapped potential both natural resources and human. It begins with authentic engagement, creating an environment for openness and collaboration among key parties. Together, plans are co-created and translated into action. Yet, the pivotal step lies in effective implementation, requiring champions to turn the vision into reality.

Now, you stand poised for action:

- Identify regions ripe for transformation and establish partnerships with suitable allies.
- Equip the team with the skills to explore new horisons, innovative thinking to co-design the future for the region and champion transformative initiatives.
- Envisage bold futures, harnessing the limitless potential of human and natural resources.
- Design system solutions to serve as the pillars of the future region's prosperity.
- Chart pathways towards building resilient and sustainable regions.

Through DSP, an ecosystem is created, paving the way for a green future.

While we have explored the application of DSP to the mining and metal industries, there is a world of untapped potential waiting to be discovered. What lies ahead? What opportunities await our pursuit? What actions must we take to propel ourselves forward? Begin with small steps, laying the ground for significant change. Embrace the journey of exploration, fueled by curiosity, determination and a commitment to learn. Remember, DSP is more than scientific discovery. It is about the collective efforts of passionate individuals like yourself.

Our goal is clear: to build sustainable, resilient regions in line with all 17 UN SDGs. But action is key. By taking the initiative to lead ourselves towards positive change, we become the driving force for transformation, both in our lives, in others and in the world around us.

Beyond being a personal endeavor, this call to action is an invitation to join a global movement towards sustainable prosperity. Across industries, governments, academics and communities, let's stand united in shaping a future brimming with possibilities.

Designing Sustainable Prosperity: Natural Resource Management for Resilient Regions,
First Edition. Edited by Doris Hiam-Galvez.

Together, we possess the potential to shape a future defined by sustainable prosperity. As we embark on this journey, let curiosity guide us, courage empower us, and commitment propel us forward. The time for change is now. Let's seize this moment, knowing that our collective efforts will light the way to a brighter, more sustainable tomorrow.

Credit: kuzzie/Adobe Stock

"In an ancient story from the Quechua people in Peru, a hummingbird fetches drops of water to help put out a great forest fire. The other animals laugh at her, but the bird replies, "I'm doing what I can"." (United Nations, 2023)

Reference

United Nations. (2023, January 16). *Ancient tale of hummingbird inspires UN World Water Day campaign*. Retrieved from United Nations : https://www.un.org/sustainabledevelopment/blog/2023/01/ancient-tale-of-hummingbird-inspires-un-world-water-day-campaign/

Index

Designing Sustainable Prosperity: Natural Resource Management for Resilient Regions,
First Edition. Edited by Doris Hiam-Galvez.
© 2024 John Wiley & Sons, Inc. Published 2024 by John Wiley & Sons, Inc.